# The Wonder of Weather

George D. Freier, Ph.D.

Gramercy Books
New York

The information in this book is true and complete to the best of our knowledge. It is offered with no guarantees on the part of the author, Fisher Books, or Gramercy Books. The author and publishers disclaim all liability in connection with use of this book.

This 1999 edition is published by Gramercy Books™, an imprint of Random House Value Publishing, Inc., 201 East 50th Street, New York, NY 10022 by arrangement with Fisher Books, Tucson, Arizona.

(Originally published under the title Weather Proverbs)

Gramercy Books™ and colophon are registered trademarks of Random House Value Publishing, Inc.

Random House
New York • Toronto • London • Sydney • Auckland
http://www.randomhouse.com/

**Library of Congress Cataloging-in-Publication Data**

Freier, G. D. (George D.)
The wonder of weather / George D. Freier.
p. cm.
Originally published: Weather proverbs. Tucson, AZ : Fisher Books, 1989.
Includes index.
ISBN 0-517-20194-1
1. Weather—Folklore. I. Freier, G.D. (George D.) Weather proverbs. II. Title.
551.63'1—dc21 97-47132
CIP

# Contents

# Dedication

Thanks to the many people throughout history who have observed the weather and passed their findings along to us in the form of weather proverbs. Each proverb is a short, salty, common-sense statement that covered the observations of many generations of people struggling with the weather. The old farmers, sea captains, sailors, and especially the American Indians who taught us so much about observing our surroundings, found that their existence and success depended on how they coped with the weather.

Many of the proverbs in this book were passed on to me by my mother who, in turn, learned them from relatives and friends. Many were brought to North America from various foreign countries to give the collection an international flavor and, thus, contribute to the melting pot of culture we find in North America.

# About the Author

George D. Freier was born on January 22, 1915, during a blizzard on an isolated farm in Pierce County, Wisconsin. His entire boyhood was spent on the farm with sufficient free time to attend a country school, the Ellsworth High School and River Falls State Teachers College. He then taught science and mathematics in the sawmill town of White Lake, Wisconsin, where he learned the lore of lumberjacks.

After three years he was invited to become a graduate student of physics at the University of Minnesota. At the time he received a Master's degree, World War II was under way. He was drafted into doing field research and development of torpedoes for the US Navy. Here he learned the lore of sea captains and sailors.

After the war he returned to the University of Minnesota and obtained a Ph.D. in nuclear physics. At that time experimental nuclear physics was done with electrostatic generators which exhibited many of the characteristics of large thunderstorm clouds.

After joining the teaching and research staff at the University of Minnesota, he switched his research efforts to atmospheric physics where he studied thunderstorms and lightning. At the time of Apollo 12 he worked with the Lightning and Transients group at Cape Kennedy to find ways for preventing lightning strikes to NASA rockets. He is continuing his work in atmospheric electricity.

George authored two previous books and many research papers. His first book was *University Physics, Experiment and Theory*. His second book, coauthored with F. J. Anderson, is entitled *A Demonstration Handbook for Physics*. This second book is widely used in American colleges and universities and has been translated into Japanese for use in schools in Japan.

# Preface

Throughout the centuries people have had to learn to cope with the weather in order to survive. They soon learned that plants and animals managed to cope with the weather through many diverse mechanisms. These mechanisms were often sensitive responses to weather changes. By watching plants and animals, people could often foresee what tomorrow's weather might be. To survive they had to observe nature much more closely than most of us now do. Their discoveries and experiences were summarized in proverbs. Tennyson described proverbs as "jewels five words long that on the stretched forefinger of all time sparkle forever." Bacon said, "The genius, wit and spirit of a nation are discovered in its proverbs."

With the advent of scientific positivism we have replaced these simpler observations with highly sophisticated digital recording and computing systems operated and interpreted by experts who then tell us about the weather via radio and television. Often they are telling us exactly what the behavior of a closely observed flower or bee can tell us. Sophisticated equipment and television often tell us the same thing we can find for ourselves from a weather proverb and careful observation of nature.

In this book you'll find a tendency to back away from positivism towards a more holistic approach to what you can learn about the weather. Using weather proverbs is certainly not as good as consultation with the Weather Service. But there are instances where communication with others is lost. We have to face the problem at hand without help. There's a lot of backup knowledge in weather proverbs.

***A king's crown does not cure a headache.***
***A doctor and a boor know more than a doctor alone.***

Rather than list every proverb uncovered in this study, I give representative samples with reasons for why we should know them and how they work.

# Proverbs

*How restless are the snorting swine,*
*The busy flies disturb the kine [cows].*
*Low o'er the grass the swallow wings,*
*The cricket, too, how sharp he sings.*
*...'Twill surely rain; I see with sorrow,*
*Our jaunt must be put off tomorrow.*

Erasmus Darwin

# 1 Why Know the Weather?

*Do business best*
*when the wind's in the west.*

A good way to understand why we should know the weather is to think about waking in the morning and facing a new day. We may feel fine or we may feel badly, depending on how well we slept. Too many or too few bed covers may reflect how isolated we are from the outside weather by protection from the blankets and house. If it is a bright, sunny morning due to high pressure in the region, we get off to a much better start than when it's damp and cloudy with low pressure.

Even before opening our eyes, our body has adjusted to atmospheric-pressure changes during the night that left our body fluids and chemistry functioning differently. We can't shield ourselves from pressure changes as we can from temperature changes by adding or removing blankets.

While washing and combing, you wonder how to dress for the day so you can live with the weather most comfortably. The first look in the mirror may be frightening if the air is humid and your hair hangs in a straight tangled mess. You may then eat a hearty or slim breakfast depending on how your body chemistry is behaving.

As you leave the house for work or play, you wonder whether or not to take boots and an umbrella. While riding to your work destination, you may wonder how the weather will be for the garden party or church social, or how your plans may be altered by a change in the weather. Your interactions with other drivers and riders may reflect how your plans are being realized.

You may be hoping that it's like this or it isn't like this on the coming weekend when you plan on going to the lake or some other outing. You may wonder how a neighbor is doing on his fishing trip to the northern part of the state, or how friends are doing on their camping trip to the mountains. At that moment, their plans are more subject to the weather than yours are.

Your thoughts may quickly turn to plans for your next vacation and you may remark, If the weather would be nice, that would be ideal. You then realize something should be done about the garden you started or plan to start when the weather continues as it is or changes.

A look at the newspaper headlines then makes you realize that people are starving in another part of the world because of drought, while elsewhere others are losing their lives and property because of storms and floods. Is the climate really changing or should we expect things like this to happen? Some governments may be toppling because their people have insufficient food and fiber. You then think that if prices keep rising at your grocery store because the farmers down state had a poor growing year, then maybe we should throw the rascals out of our state government. Why didn't the Weather Service predict this would happen so the farmers could have planted some other kind of crop? All day long your thoughts struggle with consequences of the weather in one way or another.

If you are a farmer, pilot, construction worker or taxicab driver, the weather presents many more problems that directly affect your well-being. You may survive or fail depending on how you react to the weather. If you are the president or ruler of a country, your government may stand or fall depending on how the weather is treating your subjects. Your backup military forces may even flounder if the weather is too adverse.

You will quickly conclude that weather affects your life. Some may feel content with weather information obtained from radio and television. However, you may not have access to this information at all times. And, even though you may have access to weather reports, you may prefer to proceed more independently. In this book, you'll see how weather proverbs can help with your decisions. If you retrace the above paragraphs with such sayings in mind as . . .

***When your joints all start to ache,***
***Rainy weather is at stake.***

. . . you can see how proverbs will help with your decisions involving the weather. When looking out the window after awakening we recall:

***An evening red and a morning gray,***
***Will send the traveller on his way;***
***But an evening gray and a morning red,***
***Put on your hat, or you'll wet your head.***

When looking in the mirror, we may recall:

***Curls that kink and cords that bind,***
***Signs of rain.***

When eating breakfast we may find:

***When one licks the platter clean,***
***Fair weather.***

A quick look at the skies when we leave home may make us remember that:

*Mackerel skies and mares' tails,*
*Make tall ships carry low sails.*

As we start to work, it may help to know:

*Do business best*
*when the wind's in the west.*

or

*Men work harder*
*when the pressure is high.*

These proverbs may add little or even make us more confused about the weather. Others say superstitions are entering our thoughts and we are dealing with old wives' tales. I argue that knowing only a few proverbs and knowing very little meteorology may lead to these conclusions. If, on the other hand, you know lots of proverbs and learn a small amount about how weather systems work, you may conclude that proverbs tell us much about the weather. Even if the proverbs aren't ideally accurate, they'll remind us about a weather system and help us learn and remember how weather systems work.

More important, weather proverbs make us much more aware of our environment so we can observe and appreciate the many wonderful and intricate things going on in our natural surroundings. I also think our present society likes to be spoon-fed too many things. And, we tend to get quite lazy about figuring things out for ourselves. Proverbs can help us understand and figure out various weather situations. In fact, you may find it helpful to invent some proverbs to help remember facts about the workings of the weather.

A weather proverb—a short, sensible, salty statement about the weather—is, in general, a lot of light on a subject integrated into one bright flash. The study of weather proverbs is known as *paroemieology*.

# 2 Weather Theory

***When the sun sets unhappily***
***With a red veiled face,***
***Then will the morning be***
***Angry with wind and storm.***
American Indian

When we speak of *weather,* what do we really mean? We differentiate between fair weather and foul weather, between cold and warm weather, between windy and calm weather, between sunny skies and cloudy skies, between wet and dry weather. And there are other contrasts. On the other hand, we find many of these contrasting weather descriptions fall together in a systematic way. When we know something about one set of combinations, we often know something about other combinations. For example, if the weather is wet, we very likely have cloudy skies. Knowledge about one thing seems to give information about a more general situation.

When we speak of information, we usually think of having gathered some facts. From these facts, we gain a more general understanding about a more general situation. Often, gathering information is a process of sorting contrasting statements about something. We try to find yes and no answers to simple questions, such as in the game of twenty questions. In the game, yes and no answers are related. And, as these answers accumulate, a clearer picture develops of what one is trying to discover. The game is not too different from sorting out facts about the weather. In the game and in describing weather, we add information until a satisfactory answer is found.

In understanding the weather or playing twenty questions, we are dealing with a changing situation. The common element that is changing is the *probability* of an event. As the facts about the weather are gathered, the *probability* for a future weather situation becomes more certain. Or, we say the probability approaches *unity.* Just as in the game, the probability increases as we ask questions until we find the correct answer.

The study of changing probabilities has led to a discipline known as *Information Theory.* In this discipline one can obtain a quantitative value for how much one knows. Information Theory follows many of the rules set down in Thermodynamics where one tries to evaluate the order and disorder in physical systems in terms of a quantity defined as *entropy.* Making order from disorder in physical systems allows us to derive useful work from chaotic heat energy. Making order from disorder in accumulated facts leads to knowledge. Putting observations of our surroundings in order will help us predict the weather.

The mathematics in all of these disciplines can become quite complicated, but fear not. No attempt is made to present these complicated concepts in this book. However, the concept of applying observations of our surroundings and matching them with weather proverbs should give some order to how you might predict tomorrow's weather.

# 3 How We Learn About the Weather

***Your tests show you are indecisive, procrastinating and capricious. You're hired. The Weather Service can use a person like you.***

Most of us get our weather information from television, radio or the newspaper. Weather reports cover temperature, pressure, change in pressure during the last few hours, relative humidity, and wind speed and direction. Many of us respond to this like poor people listening to stock-market reports—"just more numbers."

If we know meteorology, weather-report numbers are much more meaningful. If you have many financial investments, the stock-market reports are more meaningful. While only a few are interested in the stock market, practically all of us are concerned with the weather. This book will familiarize you with meteorology so you can watch your investment in the weather.

Television weather programs seem to come through as advertisements for the Weather Service sandwiched between other ads. But remember that they have a serious problem. They must broadcast to people distributed over a large area and do it in a very limited time. The Weather Service does an amazing job of collecting weather information and organizing it for presentation in an understandable way. The amount of information they collect and process every three hours is equivalent to writing down and sorting out the names and addresses of every person on earth. Besides this sorting job, they also must give the information a sound physical interpretation.

The Weather Service collects weather information from weather stations, including data from balloon *soundings* at only two or three stations in every state, making the average distance between stations about 200 miles. Many television stations add a denser network of observations for temperature and precipitation. Weather stations also show radar pictures and satellite pictures to give further information about situations between primary observation points.

Part of the theme of this book is to show how weather proverbs can add considerable information about a particular weather situation in an immediate surrounding and aid us in making our own predictions.

When camping, hunting or fishing, we often lose contact with radio and television information sources just as we lose contact with news reports. Because we become used to hearing the weather and news together, we get the feeling that if one of these is cut off we lose the other. Weather information does not have to cease when we're in the woods if we know how to make and use a few observations of our own. Many things in nature are more sensitive to weather than we are. If we observe the reaction of these informers to weather changes, we can learn a lot more about weather changes. Such observations of nature are often covered by proverbs. And proverbs relate observations to the weather.

In years gone by, farmers and seamen had to get all of

their weather information from their immediate surroundings. They knew a lot about what the weather might be at a later time or at a different place—often quite far away. Most weather observations were related and remembered in proverbs they knew and recited.

A good example of the many proverbs is found in the poem by Erasmus Darwin:

*The hollow wind begins to blow,*
*The clouds look black, the glass is low;*
*The soot falls down, the spaniels sleep,*
*And spiders from their cobwebs peep.*

*Last night the sun went pale to bed,*
*The full moon's halo hid her head;*
*The boding shepherd heaves a sigh,*
*For see a rainbow spans the sky.*

*The walls are damp, the ditches smell,*
*Closed is the pink-eyed pimpernel.*

*Hark, how the chairs and tables crack,*
*Old Betsy's nerves are on the wrack;*
*Loud quacks the duck, the peacocks cry,*
*The distant hills are seeming nigh.*

*How restless are the snorting swine,*
*The busy flies disturb the kine [cows];*
*Low o'er the grass the swallow wings,*
*The cricket, too, how sharp he sings.*

*Puss on the hearth with velvet paws,*
*Sits wiping o'er her whiskered jaws;*
*Through the clear stream the fishes rise*
*And nimbly catch incautious flies.*

*The glow-worms, numerous and light,*
*Illumed the dewy dell last night;*
*At dusk the squalid toad was seen,*
*Hopping and crawling o'er the green.*

*The whirling dust the wind obeys,*
*And in the rapid eddy plays;*
*The frog has changed his yellow vest,*
*And in a russet coat is dressed.*

*Though June, the air is cold and still,*
*The mellow blackbird's voice is shrill;*
*My dog so altered in his taste,*
*Quits mutton bones on grass to feast;*
*And see yon rooks, how odd their flight,*
*They imitate the gliding kite,*
*And then precipitously fall,*
*As if they felt the piercing ball.*

*'Twill surely rain; I see with sorrow,*
*Our jaunt must be put off tomorrow.*

If you like to sing, the poem goes well with the tune for *Jimmy Crack Corn.* For a chorus you may use:

***Weatherman says chance for rain,***
***Weatherman says chance for rain,***
***Weatherman says chance for rain,***
***The proverbs say the same.***

Over the years, people have noticed how many facts about the weather fit together to form what we call *weather patterns.* The Weather Service has developed many instruments for careful measurements. They put these measurements together in a scientific way to make accurate weather predictions. They feed information into sophisticated computers and, after making many calculations, make a prediction, usually by way of the local television or radio station. To make better predictions, the Weather Service would like to have many more measurements and have larger computers to analyze the data. They could then, no doubt, give us even better predictions.

Many of the adages are recalled from boyhood days before the advent of radio and television. Farmers had little more to go on than these adages to make their weather predictions. They were very observant people. Their lives depended very much on how they could cope with the weather. The more successful farmers were the ones who could predict best. And they always talked about weather in terms of weather proverbs. Farmers who knew the most proverbs seemed to be the most successful. This observation leads me to believe there is weather truth in these proverbs.

Adages are often given in verse form which makes remembering them easier. Sometimes the correct sense of a proverb can be lost in this rhyming effort, however. A good example of this has been given by Albert Lee in *Weather Wisdom* where he describes the inscription on an old tombstone:

***Underneath this pile of stones,***
***Lies what's left of Sally Jones;***
***Her name was Briggs, it was not Jones,***
***But Jones is all that rhymes with stones.***

# 4 Changing Probabilities

*The moon and the weather may change together,*
*But a change in the moon does not change the weather.*
*If we had no moon at all, and that may seem strange,*
*We shall have weather that's subject to change.*

When we change probabilities for something happening, we are placing *constraints* on the situation. What does this mean? A constraint in the case of weather means we confine our knowledge of weather development to a definite or a more ordered course of action.

A simple example of constraints may be that of driving your car down a street. If signs along the way do not allow right or left turns, no stopping is allowed and you must maintain a speed of 20 miles per hour, your course of motion is well defined. Things are well ordered as to where you will be in say 12 minutes; namely, four miles down the same road. This is providing you adhere to the constraints written on the signs. They determine exactly where you will be. A contrasting situation would be *no* speed limit or turn restrictions, no constraints. In this case, it would be impossible for anyone not controlling your vehicle to know where you will be in 12 minutes.

The example shows that the phrase "law and order" means exactly what it says. The law places constraints on what you can do, thus enhancing the probability of what will happen.

We can compare the job of a space engineer firing a rocket to the moon to that of a meteorologist trying to predict tomorrow's weather. Both predict some future event and both must contend with the laws of nature. The space engineer knows that laws of motion and gravity place sufficient constraints on the rocket's motion to know it will go to the moon with a probability of unity. The meteorologist deals with the motions of many different interacting systems. He can't know all the constraints on very complicated systems. So, not knowing all of the constraints on the system, the best that he can do is to apply those that he knows. But these are insufficient to determine the motions completely. So they can only state a probability less than unity for a certain thing to happen.

Understand that constraints can increase the probability of what can happen. The corresponding increase in probability enhances information we have about a system. Probabilities measure what we know about a system. When we know more of the constraints, we know more about what will happen.

Weather proverbs may be able to tell us something about the constraints on a weather system. Watching flowers and insects tells us something about the temperature that serves as a constraint. Watching a flight of birds may tell us something about the pressure that serves as a further constraint. By watching how tight knots are in ropes, we add another constraint. The application of many proverbs can change the probability of what we predict for tomorrow's weather.

Weather proverbs can be related to some constraint that can be placed on the weather system. If we understand a little about what a particular constraint can do to a weather system, we will be changing the probabilities and, hence, have more information about future weather. The use of weather proverbs will not yield results as good as those found by the Weather Service, but you can see they may add a little more toward making predictions.

An important point to keep in mind is: Any constraint found through the use of proverbs will apply to your situation here and now. The accurate constraints found by the Weather Service may be 100 miles away and six hours ago.

Just as golf and tennis are physical work and physical play, the use of weather proverbs can be mental work and mental play. Just as we are in better physical shape after a game of golf or tennis, we may be in better mental shape after the mental exercise of tussling with the application of weather proverbs.

There may be a time when you're boating across a big lake or hiking in the mountains that the information you gather from weather proverbs might be what differentiates between good and poor judgment. There is a proverb that says, "Poor judgment leads to experience, and experience leads to good judgment." But you can not afford to experience *everything*. Proverbs express the experiences of others.

The situation is much like the construction of chords for a musical scale. A major chord, for example, has a root note, a second note two tones higher, and a third note one and one-half tones above the second note. This is similar to the major harmony provided by the weatherman. If we add a fourth note one and one-half tones above the third note, we have a major seventh chord. Even the most inexperienced ear can tell something has been added. I hope your observations of nature, combined with your use of weather proverbs, will be this added something that changes the harmony in the world in which you live.

Each proverb can add some new tone and give a new chord. As the chords are placed together in correct sequence, a melody may develop that helps you understand the weather. Perhaps a good way to think about the whole business is to let the Weather Service, with all of its scientific observations and calculations, write the tune and lyric for the day. You will chord along with them using chords constructed from weather proverbs. If, on occasion, you lose contact with the weatherman, you can still play your chords. And from the progression of chords, you can construct your own weather melody. It may not be the best melody, but it is fun to try.

# 5 Short-range Proverbs

*Above the rest, the sun who never lies,*
*Foretells the change of weather in the skies;*
*For if he rise unwillingly to his race,*
*Or if through the mist he shoots his sullen beams,*
*Frugal of light, in loose and straggling streams,*
*Expect a drizzling day and southern rain,*
*Fatal to fruits and flocks and promised grains.*
Virgil

*Women can predict rain better than men*
*by noting how their hair stays up.*

The first group of proverbs we'll discuss are what I call *short-range proverbs.* They deal with weather for four or five days in advance. This is about the duration of time a weather system takes to move across North America. Once a weather system forms, it behaves in quite characteristic ways as it moves from west to east. This characteristic behavior is well understood by meteorologists, and they can make very accurate predictions for future weather behavior. Short-range proverbs deal with physical laws that make a weather system behave the way it does.

In each case, I give a short explanation of the physical laws that may make a proverb true. These physical laws are the basis for the behavior of a weather system. Some of these explanations have not suffered through a lot of research and are speculative. However, under expected weather conditions, we know that the physical laws stated must be operating. They will be the reason for truth, if there is any, in the proverb.

Just as "One swallow does not a summer make," so will one proverb not the weather make. You must apply many proverbs to get all the necessary information about the weather.

Proverbs from different localities or countries are often similar. In such cases, a proverb is given with an explanation of why it is likely to be correct. Then similar proverbs abiding by the same explanation are listed. Accept the one you like best.

✦ ✦ ✦

***If the goose honks high, fair weather;***
***If the goose honks low, foul weather.***

"Honking high" or "honking low" does not mean the pitch of a goose's call; it means how high the geese are flying. When we watch geese flying in their "V" formations, we realize they are marvelous engineers of flight, always adjusting to optimum airflow patterns.

They also adjust their flight to optimum air density. When barometric pressure is high, optimum density is high in the sky; when barometric pressure is low, optimum density is closer to the ground. Fair weather occurs during times of high pressure; foul weather occurs during time of low pressure. The flight of geese correlates well with barometric pressure.

***The low flight of rooks indicates rain.***

***Birds flying low,***
***Expect rain and a blow.***

***If the lark flies high,***
***Expect fair weather.***

***When swallows fleet soar high and sport in the air,***
***He told us that the welkin would be clear.***

***Wild geese fly high in pleasant weather***
***And fly low in bad weather.***

***Everything is lovely when the goose honks high.***

✦ ✦ ✦

***If the rooster crows on going to bed,***
***you may rise with a watery head.***

This proverb predicts rain. Roosters, like other birds, crow to declare territorial rights, usually in the morning. When a rooster crows in the evening, he has somehow developed a more quarrelsome mood and feels more like picking a fight. Like most animals, he is experiencing a pressure change due to a falling barometer. His body must be giving up dissolved gases as he adjusts to falling pressure. He must do this or he would tend to explode. When body fluids give up dissolved gases, air molecules are not released molecule by molecule, but tend to nucleate into bubbles. These small bubbles tend to affect nerve impulses at synapses so the animal is more irritable. This is true for people, too.

***If the raven crows, expect rain.***

***When geese cackle, it will rain.***

***When ducks quack loudly, it's a sign of rain.***

***The hooting of the owl brings rain.***

***If the sparrow makes a lot of noise,***
***rain will follow.***

***When parrots whistle, expect rain.***

Fish start biting when the barometer starts to fall. Part of rainy weather is the development of a low-pressure center. At all times, there is decaying plant matter at the bottom of lakes and streams. Part of the decay products is gas that forms small bubbles that cling to the decaying matter. Many insect nymphs also reside in the decaying matter.

*Fish bite best before a rain.*

When barometric pressure starts to fall, the bubbles expand and make some of the decaying matter sufficiently buoyant to rise. Minnows start to feed on this rising matter, and larger fish start feeding on the minnows to set the whole feeding chain in progress.

***When fish break water and bite eagerly, expect rain.***

***When porpoises sport and play,***
***there will be a storm.***

***Trout jump high***
***When a rain is nigh.***

***Before the storm***
***The crab his briny home forsakes,***
***And strives on land to roam.***
Aratus

✦ ✦ ✦

***Bubbles over calm beds of water mean rain is coming.***

Bubbles come from gases generated at the bottom of a lake. As water plants die and sink to the bottom, they continue to decay. Part of the decay products is gas which normally clings to the decaying material. When barometric pressure falls, as it usually does before a rain, bubbles expand and become more buoyant. With this added buoyancy, the bubbles can break loose from decaying matter and rise to the surface. With an excess number of bubbles at the surface, we know that barometric air pressure is falling and conditions are getting favorable for rain. Often there is foam on the surface and the water is murky. Some of the decaying material brought to the surface is phosphorescent.

***Marshes give off an eerie light before a rain.***

***Wells give murky water before a storm.***

***Look for foam on the river before a rain.***

***When bubbles are rising on the surface of coffee and***
***they hold together, good weather is coming;***
***If the bubbles break up,***
***weather you don't need is coming.***

✦ ✦ ✦

***Underground miners can smell rain coming.***

This was probably more true in olden days than it is today when mines have better ventilation. It is also more likely to happen in coal mines. We know that coal is decayed vegetable matter and, in this decayed matter, there are smelly gases such as methane. When the barometer falls, these trapped gases diffuse to the lower pressure in the mine where the miners can smell them. Much worse than the bad smell is the possibility of developing an explosive mixture. Often there is carbon monoxide in the gas. Many people have lost their lives by simply being in a deep hole in the ground while barometric pressure falls.

A falling barometer means a low-pressure system is moving in, making the outside air unstable. If there is sufficient moisture in the air, the rising air will produce rain.

*When ditches and ponds*
*Offend the nose,*
*Look for rain*
*And stormy blows.*

*When boiling water more rapidly vanishes, expect rain.*

*If pavements appear rusty, rain will follow.*

✦ ✦ ✦

*When the glass falls low,*
*Prepare for a blow;*
*When the glass is high,*
*Let your kites fly.*

The glass was an early settler's barometer. It was a closed tube at one end and open at the other, filled with water, and a bubble of air trapped above the water at the closed end. Water level at the closed end would rise or fall with air pressure. It was not very accurate because the air bubble would dissolve in the water and the water would evaporate. Barometers are now made with mercury, which is very dense, and no air bubble is needed. Although very expensive, mercury evaporates very slowly. A column of mercury is the official pressure-measuring device of the Weather Service. The proverb tells us weather is fair when pressure is high, foul when it's low.

*In the winter, a heavy snow is predicted if the barometer falls and the temperature rises.*

*When the wind backs, and the weather glass falls,*
*Then be on your guard against rain and squalls.*

*If cirrus clouds form in weather with a falling barometer, it is almost sure to rain.*

*A summer thunderstorm that does not depress the barometer will be very local and of little consequence.*

*Springs start to flow just before a rain.*

Springs occur where the ground-water table reaches the surface. Deep in the water, total pressure is due to: (1) weight of the water and (2) atmospheric pressure. Deep points all tend to come to the same pressure. If the barometer falls in one region more than in another at some large distance, the smaller atmospheric pressure will be compensated for by higher water pressure or rising water levels. Springs will then start to flow where the barometer is falling. The falling barometer is accompanied by unstable air, both necessary conditions for rain. This effect is much more marked over oceans when hurricanes come. The very low barometer readings in a hurricane are accompanied by rising water or very high tides.

*Sap from the maple tree flows faster before a rain shower.*

*Water rising in springs and wells indicates rain.*

*Many springs that have gone dry will have a good flow of water before rain.*

*Wells gurgle and yield muddy water before a storm.*

✦ ✦ ✦

*Soot falls down before a rain.*

Soot is mostly carbon particles that are very porous. They can readily absorb gases and store them in microscopic cells within the particle. During high barometric-pressure periods, the cells become filled with gas at relatively high pressure. When the barometric pressure starts to fall, some of the gas must escape. If the gas cannot escape sufficiently fast, each cell may rupture and break away from its support place in the chimney. The falling-soot problem is also aided by higher humidity.

Water from the air can condense in the channels of very small radius. These channels are open during drier high-pressure periods when air enters the cells, but become plugged by condensing water as humidity increases. Falling pressure and high humidity are both forerunners of rain.

***If burning coals stick to the bottom of a pot,***
***it is the sign of a tempest.***

***Fires burning paler than usual and murmuring within***
***are significant of storms.***

Bacon

***Burning wood pops more before rain and snow.***

✦ ✦ ✦

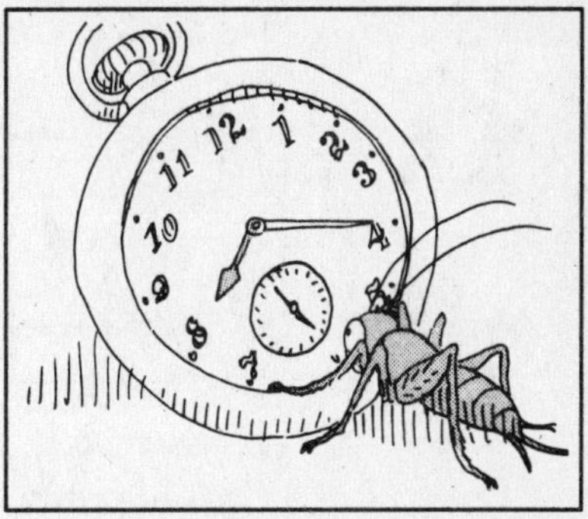


***Count the number of cricket chirps in 14 seconds, add 40, and you have the temperature in degrees Fahrenheit.***

Crickets are cold-blooded animals, meaning their body temperature is always at the temperature of the surroundings. Their activity is closely related to ambient temperature. They chirp by sawing on their bodies with their back legs. The higher the temperature, the faster they saw. If we listen to many crickets chirping, we find that they chirp in unison, all indicating the same temperature. Students in science projects have tested this quite thoroughly. Some insects respond to temperature changes by the duration of their "song" rather than the rate. Long songs indicate high temperature.

*The katydid's song gives the following temperatures:*

| | |
|---|---|
| *Kay-tee-did it* | *78F* |
| *Kay-tee-didn't* | *74F* |
| *Kay-tee-did* | *70F* |
| *Kate-Didn't* | *66F* |
| *Kate-tee* | *62F* |
| *Kate* | *58F* |

*Cockroaches are more active before a storm.*

*Locusts sing when the air is hot and dry.*

*Ants are very busy, gnats bite;*
*Crickets are lively, spiders leave their nest;*
*And flies gather in houses before a rain.*

✦ ✦ ✦

*Open crocus, warm weather;*
*Closed crocus, cold weather.*

The crocus is a very temperature-sensitive flower. Botanists say you can tell the temperature to the closest one-half degree by noting how far the flower has opened. The tulip is another flower that's quite temperature-sensitive. If, on a warm spring day, you pick a tulip that is wide open, then place it in a refrigerator, the flower will close into a tight bud. Upon being removed from the refrigerator, the flower opens wide again.

***Tulips open their blossoms when the temperature rises, they close again when the temperature falls.***

***The daisy shuts its eye before rain.***

***If the marigold should open at six or seven in the morning and not close until four in the afternoon, we may reckon on settled weather.***

✦ ✦ ✦

***When pipes smell stronger, it's going to rain.***

In addition to smoke particles, which are visible, many complex molecules are given to the air which are invisible but pungent. These molecules are more or less "naked" in dry air. But, in moist air, water molecules in the air tend to collect on the aromatic molecules, covering them with a layer of water molecules. We say the large aromatic molecules become *hydrated.* This layer of water molecules then allows aromatic molecules to bind better to the moist surfaces in your nose where the sense of smell is located. With more water molecules in the air, relative humidity is increasing, making rain more likely. Not all smells need be bad. A similar thing happens for the aromatic molecules from flowers. They smell better when the air is moist.

***If the perfume of flowers is unusually perceptible, Expect rain.***

***Flowers smell best just before a rain.***

***When ditches and ponds offend the nose, Look for rains and stormy blows.***

*Manure piles smell stronger before a rain.*

✦ ✦ ✦

*If cats lick themselves, fair weather.*

During fair weather, when the relative humidity is low, electrostatic charges (static electricity) can build up on a cat as it touches other objects. Cat hair loses electrons rather easily, so cats becomes positively charged. When a cat licks itself, the moisture makes its fur more conductive so the charge can "leak" off the cat. In fair weather during high pressure, air is sinking from above. This air is quite dry, having its moisture wrung out in other places. Relative humidity is then low and things such as cat hair become better insulators. Many cats do not like to be petted during cold winter weather when the humidity is very low because sufficient charge builds up to cause small sparks which irritate them. The original definition of negative charge was that charge transferred to an ebonite rod from cats' fur.

*If a dog pulls his feet up high while walking,*
*A change in the weather is coming.*

*Cats scratch a post before wind,*
*Wash their faces before rain,*
*And sit with backs to the fire before snow.*

*Cats with their tails up and hair apparently electrified indicate approaching wind.*

✦ ✦ ✦

*If spiders are many, and spinning their webs, the weather is fair.*

Spider webs are very sensitive to the amount of moisture in the air. When relative humidity is high, indicating much moisture in the air, spider webs absorb moisture. The threads become thick and tight to the extent that many threads will break. When a spider builds his web in dry fair weather, the web tends to self-destruct as soon as the air becomes more moist. On a dry fall day, you may see *gossamer* in the air which is many spider webs with small spiders attached. Small spiders migrate by means of "parachutes" blown in the wind. They spin a web and then launch themselves into the breeze.

In fair weather during high barometric pressure, the air descending from above is dry because the moisture has condensed out as rain in other regions.

*When spider webs in air do fly,*
*The spell will soon be very dry.*

*If garden spiders forsake their webs,*
*It indicates rain.*

*If spiders are many and spinning their webs,*
*The spell will soon be very dry.*

*Spiders enlarge and repair their webs before bad weather.*
American Indian

✦ ✦ ✦

*A reddish sun has water in his eye;*
*before long you won't be dry.*

A reddish sun can portray either wet air or dry air. Dry air is a more pinkish red; wet air is a deeper red. This proverb deals with the deeper red. All air contains many small particles too small to scatter light and too small to be seen with the unaided eye. When air becomes more moist, water molecules in the air collect onto small particles to make them larger. The sun gives us white light which includes all colors from short-wavelength blue to long-wavelength red light.

Water particles start increasing in size as they add water molecules. They first scatter the short-wavelength blue light and leave the remaining light from the sun more red. This proverb says the air is becoming more moisture-laden, a necessary condition for rain. It is not a sufficient condition by itself because other things must happen before the enlarging drops can continue to grow to the size of falling raindrops. Pinkish red is observed when dry winds raise a sufficient amount of dust so some blue light can be scattered without the addition of water molecules. The air is too dry, in this case, for further growth by adding water molecules.

*When the sun sets bright and clear,*
*An easterly wind you need not fear.*

*Clouds on the setting sun's brow indicate rain.*

*Evening red and morning gray;*
*A good sign for a fair day.*

*When the sun sets unhappily*
*With a red veiled face;*
*Then will the morning be*
*Angry with wind and storm.*
American Indian

*A red evening and a gray morning*
*Sets the pilgrim walking.*

*An evening red, and a morning gray,*
*Sets the traveler on his way;*
*But an evening gray and a morning red,*
*Put on your hat, you'll wet your head.*

*If the sun in red should set,*
*The next day surely will be wet;*
*If the sun should set in gray,*
*The next will be a fair day.*

*An evening gray, and a morning red,*
*Makes the shepherd hang his head.*

✦ ✦ ✦

***When walls are wet,***
***expect some rain.***

The walls referred to in this proverb are usually stone or masonry. Rocks, in general, have high heat capacity. This means it takes a great deal of heat to make their temperature rise very much. Sunshine can warm the air, plants and other surrounding objects faster than it can raise the temperature of a stone wall.

If the air is quite moist, the rocks can remain at a temperature below the dew point while the surrounding air is at a temperature above the dew point. If the air is quite moist, the dew-point temperature is only slightly below the real air temperature and moisture can condense on the cool rocks.

If the air is very dry, the dew-point temperature is much below the real temperature so that the rocks, which are only slightly below the real temperature, are above the dew-point temperature and remain dry. In Finland, people keep a type of smooth stone outside their homes and watch it carefully to help predict the weather.

*If metal plates and dishes sweat,*
*It is a sign of rain.*

*Quarries of stone and slate indicate rain by*
*moist exhalation from the stone.*

*When stones sweat,*
*Rain you'll get.*

✦ ✦ ✦

*Pale moon rains,*
*Red moon blows;*
*White moon neither*
*Rains or blows.*

The moon reflects light from the sun. Because the moon has no atmosphere to absorb or scatter any of this light, changes in the moon's color come from interactions in our atmosphere. Most of the changes can be seen in a rising moon when it is full. Light from the moon must then come through the longest path of our atmosphere. In a clean, dry atmosphere, most all colors come through equally well and we see a white moon.

The air always contains condensation nuclei, but they are so small that they will not scatter light. When the air starts getting moist, condensation nuclei collect water molecules and grow. They start scattering the short-wavelength blue light, which makes the moon appear to be redder. This moisture in the air can then lead to rain. A pale moon has its light scattered by larger ice particles in the upper atmosphere and all colors are scattered to make the intensity less or the moon appear to be pale.

*The moon, her face be red,*
*Of water she speaks.*
American Indian

*If the moon rises clear, expect fair weather.*

*When the moon rises red and appears large, with clouds, expect rain in twelve hours.*

*When the moon is darkest near the horizon, expect rain.*

✦ ✦ ✦

*If salt is sticky,*
*And gains in weight;*
*It will rain*
*Before too late.*

Salt is what we call a *hygroscopic* material. This means it can take moisture from the air and retain the moisture. Some of the salt crystals then become covered with a salt solution which tends to lower the pressure of water vapor over the salt. This means there are fewer water molecules in the air next to the salt. Surrounding moist air will then tend to send in more water molecules into the region around the salt. The salt then grabs it to become heavy with absorbed moisture. The moister the air, the faster this process can proceed to make the salt gain weight and become sticky with the salt solution around the crystals. Moist air is necessary to form clouds and rain.

*When cheese salt is soft, expect rain.*

*Tobacco gets moist before a rain.*

*Soap gets slippery before a rain.*

*Oily floors quite slippery get,*
*Before the rain makes everything wet.*

✦ ✦ ✦

***If there is dew on the grass in the morning, fair weather.***

Dew forms on grass when the grass gets colder than the dew-point temperature. Grass gets cold because during the night the grass and surrounding soil radiate heat energy to a clear sky. On its way through the atmosphere, radiant energy from the grass to outer space may or may not encounter water molecules in the air.

If there is lots of water vapor in the air, it will absorb the radiation and subsequently radiate some of this energy back to the ground where it is reabsorbed at the earth's surface. This tends to keep the grass warm and above the dew-point temperature. If, on the other hand, there is little moisture in the air, radiant energy gets through the atmosphere and escapes to outer space. None is then returned to the grass, so the grass cools below the dew-point temperature to show the formation of dew. Dry air is characteristic of high-pressure, fair-weather systems in which dry air is descending from above with little chance for rain and clouds. Radiation cooling and fair weather go hand in hand.

***When dew is on the grass,***
***Rain will never come to pass.***

***Mist rising from the pond,***
***Fair weather tomorrow.***

***Clear moon, frost soon.***

✦ ✦ ✦

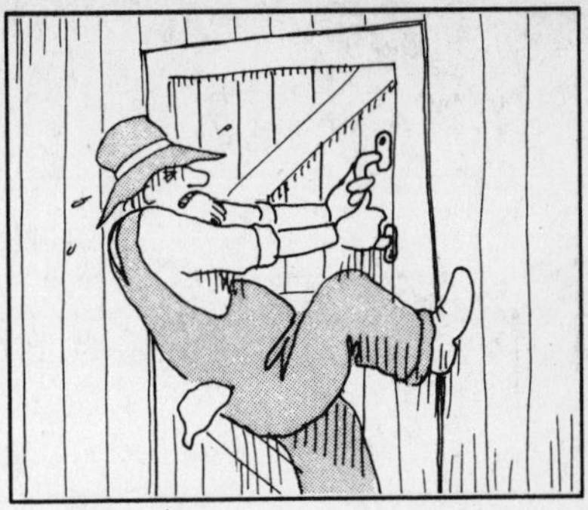

*When doors and windows start to stick, it will probably rain.*

Sticky doors and windows indicate relative humidity is high. Cellulose fibers in wood are long tubular structures that carry fluids within a tree. The diameters of these tubes are very small, so liquid water can condense in them. The water then makes wood swell or enlarge and everything fits more tightly. Water condenses in wood because liquid surfaces in the small tubes that are very highly curved lower the vapor pressure as water molecules pass to liquid form. More molecules can then move in to repeat the process. Scientists often use small capillary tubes to liquefy strange vapors. Increasing humidity is a condition under which you can expect rain.

***When mountain moss is soft and limpid,***
***Expect rain.***

***When corn fodder stands all dry and crisp,***
***Go on your outing, there's no great risk.***

***Doors and drawers stick before a rain.***

✦ ✦ ✦

Older-type ropes were made from plant fiber as many still are. The cellulose fibers from which rope is made are actually long capillary tubes that can hold water. The curved surface of any liquid water in these tubes has a very small radius of curvature that is concave to the air side. This curvature makes it easier for the liquid to catch and retain water molecules, so we say it *lowers the vapor pressure over the water surface.*

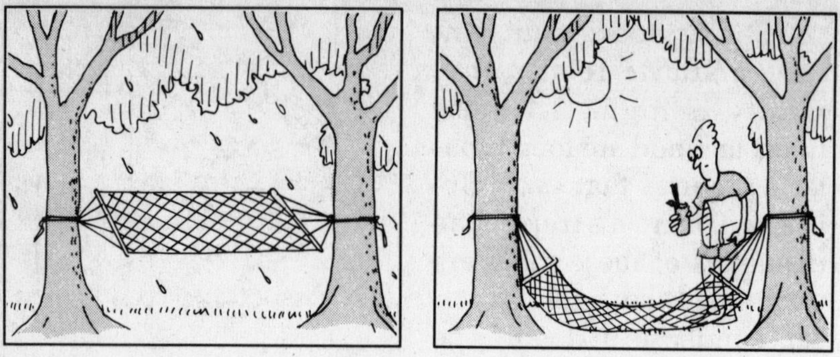

***When ropes are tight, it's going to rain;***
***When weather's fair, they're slack again.***

Water molecules in the air then enter the tubes and condense long before the outside dew-point temperature is reached. Liquid water is deposited in the rope as soon as the relative humidity starts to climb. Liquid water in the cellulose cells seems to make rope fibers expand horizontally and contract in the long direction to make ropes fatter and tighter. When the relative humidity is high, the probability of rain is greater.

***Knots get tighter before a rain.***
***Ropes shorten before a rain.***

***Sailors note the tightening of ropes on a ship before rain.***

***When locks turn damp in the scalp house,***
***It will surely rain.***
American Indian

***Guitar strings shorten before a rain.***

✦ ✦ ✦

***Dandelion blossoms close before a rain.***

Apparently a dandelion blossom can sense relative humidity in the air. The blossom is making fine silky fibers attached to seeds so the seeds can be distributed by the wind.

These silky fibers should not get wet if they are to serve their purpose. Plant cells in the pedestal of the flower can become more or less turgid and change the shape of the blossom's cover. An interesting experiment to show how plants can do this is to pick a dandelion flower, then split the end into four parts for a distance of about one inch. Dip the split end into a glass of water and remove it. In a short time, the four split-end sections will curl outward as shown in the drawing, page 109.

When the air gets more moist, there is an enhanced probability for rain. Some plants find it an advantage to open more to collect water.

***When the milkweed closes its pod, expect rain.***

***The pitcher plant opens wider before a rain.***

***Chickweeds close their leaves before a rain.***

***Closed is the pinkeyed pimpernel before rain.***

***When corn fodder is crisp, fair weather;***
***When corn fodder is limp, rain is coming.***

✦ ✦ ✦

Smoke is very small particulate matter. Often there are many more particles than we can see because they are too small to scatter light. When relative humidity is high, the particles tend to collect water molecules from the air and become heavier. Being heavier they fall to the ground.

There is also lots of water in the exhaust of a fire that collects on the particles.

***If smoke falls to the ground, it is likely to rain.***

If the air is dry, this exhaust water evaporates from the particles, making them lighter and also invisible.

The persistence of smoke is also a sign of higher humidity. Cigar and cigarette smokers can observe this as they exhale smoke. Coming from warm, moist lungs, the smoke particles are covered with water molecules and are large enough to scatter all light, so the smoke is gray.

As water evaporates from the particles, smoke turns blue because they become sufficiently small to scatter only blue light. As evaporation continues, the particles get too small to scatter any light, and thus become invisible. If the smoke persists, relative humidity is high and rain is more likely.

***Campfires are more smoky before a rain.***

***The factory smoke stack is more of a nuisance before a rain.***

✦ ✦ ✦

***Frogs croaking in the lagoon, means that rain will come real soon.***

Frogs are cold-blooded animals so the temperature must be reasonably high before they become very active. They are also aquatic animals, so their skin must be moist. Their croaking activity then implies a rather high temperature and high humidity. These conditions are both favorable for rain. If an instability in the atmosphere develops, the warm air can rise to form large clouds. Look also for signs of a falling barometer, which would further imply an approaching cold front. As the cold front passed, it would aid in lifting the warmer air and getting things started.

*Frogs croak before a rain;*
*But in the sun are quiet again.*

*If toads appear in large numbers, expect rain.*

*If frogs make a noise at the time of cold rain,*
*Warm dry weather will follow.*

*If many earthworms appear, rain will follow.*

✦ ✦ ✦

Mushrooms are most often grown in old caves where it is easier to maintain the relative humidity. When the air is humid and the atmosphere becomes unstable, we can expect rain.

*If you see toadstools in the morning, expect rain by evening.*

*When frogs jump across the road, they are looking for rain.*

*Dead branches falling in calm weather indicates rain.*

*Mushrooms and toadstools are plentiful before rain.*

✦ ✦ ✦

Towering clouds imply that the air is both rich in moisture and is becoming unstable. Unstable air is rising air. It rises because the cloud air is warmer than the surrounding air. Warm air with less density is buoyed upward by the cooler surrounding air. Cloud air is warmer because the condensing vapor is giving up its heat of condensation. This heat was originally given to the vapor when the sunshine

*When clouds appear*
*Like rocks and towers,*
*The Earth's refreshed*
*With frequent showers.*

caused the water to evaporate. The towering implies that the instability persists to great heights so more and more moist air is being pulled into the cloud at the cloud base. As the air continues to rise, more and more of it condenses until coalescence of the little cloud drops starts to give us larger drops which can become rain. We often forget how nature has such a beautiful system for collecting water and solar energy over the oceans and delivering it to faraway regions of land.

***If cumulus clouds are smaller at sunset than at noon,***
***expect fair weather.***

***When cumulus clouds become heaped to leeward***
***during a strong wind at sunset,***
***thunder may be expected during the night.***

***If wooly fleeces spread the heavenly way,***
***be sure no rain disturbs the summer day.***

✦ ✦ ✦

The following proverb refers to the heights of cloud bases. If a cloud base is low so it sinks below a hill top, we can be sure the air is quite moist. When the cloud base is at a low level, the dew-point temperature and real temperature are close together and make the air quite humid. As air rises and cools, the real temperature falls and becomes equal to the dew-point temperature, so condensation starts and clouds form.

If cloud bases are higher than the hills, the dew-point temperature and real temperature are far apart at the ground, indicating dry air. The air must then rise much higher before the real temperature equals dew-point temperature, so the cloud bases are high. Rising cloud air, which draws moist air in at the base, is much more likely to produce rain.

*When clouds sink below the hills, foul weather;*
*When clouds rise above the hills, fair weather.*

*After black clouds, fair weather.*

*If cloudy and it soon decreases,*
*Certain fair weather.*

*A round-topped cloud and flattened base,*
*Carries rainfall in its face.*

*When mountains and cliffs in the clouds appear,*
*Some sudden and violent showers appear.*

*Long foretold, long last*
*Short notice, soon passed.*

✦ ✦ ✦

*Enough blue sky in the Northwest to make*
*a pair of Dutchman's breeches is a sign*
*of approaching fair weather.*

Storm clouds are always tall clouds, so there is sufficient height for precipitation to develop. Low clouds are more or less just foggy air raised above the ground. If you can see blue sky through the clouds, you know the high clouds are vanishing and with them the possibility of precipitation. The sun can then furnish heat to the low clouds and burn them away much as it burns away the morning fog. Usually this situation is associated with a cold front passing, which produces stormy weather as it passes. In general, fronts move from northwest to southeast. Seeing blue sky in the northwest means the front has almost passed, being followed by cooler, drier air with clear skies and fair weather.

*When clouds are upon the hills,*
*They'll come down by the mills.*

*When Lookout Mountain has its cap on,*
*It will rain in six hours.*

*Clouds upon the hills, if rising, do not bring rain,*
*If falling, rain follows.*

*If clouds rise in heaps of white,*
*Soon will the country of the corn priests*
*Be pierced with arrows of rain.*
American Indian

✦ ✦ ✦

*Cumulus clouds in a clear blue sky,*
*it will likely rain.*

Cumulus clouds are white or gray and have well-rounded edges. Clouds indicate there is sufficient moisture in the air for condensation as the air rises and cools. The question is whether they will rise sufficiently high to start coalescence of the small cloud droplets and lead to rain. Blue sky tells us they probably can. If the sky is hazy blue instead of deep blue, it indicates there is a temperature inversion above to which air can rise and then become stable. At this stable layer, upward motion is arrested and clouds can not break through it. Haziness is caused by atmospheric debris such as pollution collecting at the inversion level. A deep-blue sky indicates no inversion and, hence, no limit to cloud growth. The clouds can continue to grow and eventually produce rain.

***Sunshiny shower***
***Lasts half an hour.***

***Clouds small and round like a dapply-gray,***
***With north wind, fair for a day.***

***The higher the clouds, the fairer the weather.***

***When smoke rises but not too high,***
***Clouds won't grow and you'll keep dry.***

✦ ✦ ✦

***Mackerel skies and mares' tails***
***Make tall ships carry low sails.***

Mackerel clouds are usually alto-cumulus clouds while mares' tails are cirrus clouds. Cirrus clouds are at heights of about 10 miles while alto-cumulus clouds are six to eight miles above the surface. This situation is characteristic of an approaching warm front. In this case, you are in cool air while the westward motion of weather systems is bringing in warmer, moist air from the west. Warm air, being more buoyant than cool air, is forced to rise and override the cool air. Moisture in the warm air starts to condense as it rises and forms clouds. The cold air is in the form of a large flat bubble whose edge can be hundreds of miles away.

The first you see of the warm, cloudy air is high cirrus clouds or mares' tails. As the system approaches, you see lower alto-cumulus or mackerel clouds. As the system continues to approach, the clouds get thicker until they start to rain. The slope of the cold-air bubble and the speed at which the front approaches allow the proverb to describe weather 24 to 36 hours in advance.

*Hen's scarts and filly tails*
*Make lofty ships carry low sails.*

*If the upper currents of clouds come from the Northwest in the morning, a fair day follows.*

*If cirrus clouds dissolve and appear to vanish,*
*It is an indication of fine weather.*

✦ ✦ ✦

*A ring around the sun or moon,*
*Means that rain will come real soon.*

The ring is caused by light from the sun or moon passing through ice crystals in the upper regions of the atmosphere. The ice crystals are six-sided crystals that refract the light coming toward you, which would otherwise go elsewhere. Only those crystals which are oriented to give minimum deviation can enhance the intensity of the light at the position of the ring.

Because people are all in different places, no two people see the same ice crystals. That part of the environment is made only for you. Ice crystals have been blown from the tops of high thunderstorm clouds or they come from warmer, moist air being forced over colder air in a warm-front situation. Weather systems move from west to east, so the ring implies moist, stormy air is coming your way.

*When around the moon there be a bruh [halo],*
*The weather will be cold and rough.*

*A blur or haziness about the sun*
*Indicates a storm.*

*A solar halo indicates bad weather.*

*The circle of the moon never filled a pond.*

*The circle of the sun wets the shepherd.*

*If the moon in a house be,*
*Cloud it will, rain will follow.*
American Indian

*The moon with a circle brings water in her beak.*

✦ ✦ ✦

Bees don't get caught in the rain because they stay close to their hives before rain. A bee requires some kind of navigation system to travel to distant places and return. Bees sense their orientation with respect to polarized light from the sky. In clear weather, the blue scattered light from the sky is quite polarized. That is, there is a definite plane of vibration for the light waves.

*Bees never get caught in a rain.*

Bees can sense this polarization and use it for navigation. Before a rain, there are often high cirrus clouds which are often tops from tall thunderstorm clouds to the west or condensation from overriding warmer air at a warm front. Ice crystals in these cirrus clouds destroy the polarization. The bees then lose their navigation systems and stay close to the hive. Beekeepers note a poor yield of honey during exceptionally cloudy summers.

*Bees will not swarm before a rain.*

*If bees remain in their hives*
*or fly a short distance,*
*expect rain.*

*If bees to distance wing their flight,*
*Days are warm and skies are bright;*
*When their flight ends close to home,*
*Stormy weather's sure to come.*

✦ ✦ ✦

*Rainbow in the morning*
*Sailor's warning;*
*Rainbow at night*
*Sailor's delight.*

Rainbows are caused by refraction of sunlight in each water drop and reflection of this refracted light from the backside of the drop. This is why rainbows are always on the opposite side of us from the sun. In the morning rainbows are in the west; in the evening they are in the east. A rainbow will always be at the position of the raindrops. A rainbow in the morning means raindrops to the west; a rainbow at night means raindrops to the east. Weather systems move from west to east, so a rainbow in the morning means rain is moving toward us; at night rain is moving away from us. Sailors prefer fair weather that follows rain.

*If there's a rainbow in the eve,*
*It will rain and leave;*
*If there's a rainbow in the morrow,*
*It will neither lend or borrow.*

*Rainbow to windward,*
*Foul fall the day;*
*Rainbow to leeward,*
*Damp runs away.*

✦ ✦ ✦

*Pigs carry sticks and straw before rain.*

It is difficult to analyze this proverb, perhaps because I don't know much about pigs. If it is going to rain, barometric pressure is most likely falling. Animals such as horses and cows—and even children and adults—become irritated when the pressure falls. When external pressure falls, the body must release absorbed gases or it will tend to explode.

Absorbed gases are in body fluids, but they can not be released molecule by molecule. Instead, the molecules of absorbed gas nucleate into little bubbles in the body fluids. These bubbles must somehow interfere with the transmission of nerve pulses at a synapse to make the animal not feel so well. I don't know why a pig exhibits ill feelings by carrying straw.

*It will rain when pigs scratch themselves on a post.*

*Before a rain, sheep are frisky and box each other.*

*Horses are startled and nervous before a storm.*

*When a dog rolls on his back, it will soon rain.*

*When the ass begins to bray,*
*Surely rain will come that day.*

*When horses are restless and paw with their hoof,*
*You'll soon hear the patter of rain on your roof.*

✦ ✦ ✦

*If cows huddle, it's going to rain.*

It is certainly true that cows huddle during a rain storm. Quite often an entire huddled group will be killed by lightning as they gather under a tree. Lightning strikes the tree and, if it is not well grounded, the charge spreads laterally at the base of the tree. Part of the spreading lightning current can run up one leg of a cow and down the other, electrocuting the cow. Because there are many branches of the spreading current, many cows may be electrocuted. Sometimes lightning can strike a fence at quite some distance, then jump to the tree.

Cows are *ruminants.* It seems their early survival depended on hurriedly eating lots of grass, then hiding in the bushes away from carnivores while chewing their cud for better digestion. I think the rising temperature before a rain makes them rather lethargic, so they retire to huddles early to chew their cud.

***Rabbits leave the field and***
***head for the wood before a rain.***

***Mountain goats come to lower ground before a rain.***

***Deer move to lower wooded areas before a rain.***

***When a cow will scratch her ear,***
***It will mean a rain is near.***

***Mice will run and frolic before a storm.***

✦ ✦ ✦

My experience is that the heifer has her tail in the air and runs wildly when chased by large botflies. These flies lay eggs under the more tender skin of a heifer where large maggots develop. These flies are sometimes called *green-*

*bottle* flies. When the maggot grows, a large bump about the size of a bottle neck develops. One can then place a bottle opening over the bump, press down on the bottle, and out pops the maggot into the bottle. There are several proverbs about flies biting more viciously before a storm. Usually it is quite warm and the fly, a cold-blooded insect, becomes sufficiently active to chase the heifer and lay the eggs.

***Watch the heifer's tail; when stretched aloft, 'twill rain or hail.***

***Bats flying late in the evening, indicates fair weather.***

***Bats who speak flying, tell of rain tomorrow.***

***When the cows come home with hay pieces dropping out of their mouths, then rain will come.***

***When an old cow raises her head high and sniffs the air, soon a change to nasty weather will come.***

✦ ✦ ✦

***If the robin sings loudly from the topmost of trees, expect a storm.***

Birds don't sing because they are happy but to declare their territorial rights. When barometric pressure starts to fall, as it does before a rain, most animals seem to experience some irritation as escaping gases from their bodies form little bubbles in body fluids. Feeling somewhat irritated, a bird declares territorial rights and is willing to fight for them. A second thing the robin may sense is the instability of the air which is developing turbulent motions.

Experiments with birds indicate they can hear much lower frequencies than we can. A law of physics says hearing mechanisms must have sizes somewhat close to the wavelength of the sound being heard. Because low frequencies have *long* wavelengths and birds' ears are so *small,* they most likely sense low-frequency sounds with their feathers. Feathers would make much longer sensors to match the longer wavelength of low-frequency sound. Turbulent motions in a developing storm put out these low frequencies that birds can hear, even though we can't.

***Guinea hens squawk more than usual before a rain.***

***When the peacock loudly bawls,***
***Soon you'll have both rain and squalls.***

***The blackbird's call is more shrill before a rain.***

***When cranes make a great noise or scream,***
***Expect rain.***

***When chimney swallows circle and call, a sign of rain.***

***If crows make much noise and fly round and round,***
***A sign of rain.***

***Gulls will soar aloft and circling around,***
***Utter shrill cries before a rain.***

***If the turkey's feathers are ruffled, it will rain.***

***If a hawk flies to the top of a tall tree and***
***searches for lice, it's a sign of rain.***

***When birds oil their feathers, expect rain.***

***If robins sing loud, and robins sing long,***
***It's a sign of rain.***

*Rain doves coo before a rain.*

*Cranes soaring aloft and quietly in the air is a sign of fair weather; but if they make much noise, as if consulting which way to go, it foreshadows a storm that is near at hand.*

*When birds huddle at the top of a chimney top,*
*It is a sign of cold weather.*

*Partridges perching high in a tree*
*Indicate that rain is coming.*

*Birds sitting on a telephone line, expect rain.*

*Pigeons stay close to their quarters before rain.*

✦ ✦ ✦

*Wild geese, wild geese*
*Going out to sea*
*All fine weather*
*It will be.*

Wild geese are great navigators as they fly in their "V" formations. They sense the very low frequency of storms of down to one tenth of a cycle per second. Such noise is due to turbulence we can't hear and is characteristic of developing storms. When migrating from the Atlantic seacoast, geese wait for an approaching cold front before starting their journey over water. Air-traffic controllers often see flocks of geese on their radar screens. They note that geese always fly to avoid a storm.

If birds can hear very low frequencies that we can't, they most likely hear with their feathers. These are anchored into the deeper skin dermis with many nerve endings rather than being part of the epidermis as are scales on a reptile. Migrating birds always seem to head for fair weather.

*Sea gulls stay on land before a storm.*

*Don't cut hay when the robin's in the bush.*

*Robins in the bush, rain is coming.*

*When birds stop singing,*
*A storm is on the way.*

*Hawks hunt small animals as they move to higher ground before a rain.*

*Migrating birds fly to avoid a storm.*

✦ ✦ ✦

*When the North Star starts to twinkle, it will probably bring rain.*

Stars are point sources of light in the sky, meaning light comes to us as if it were coming from a point rather than an extended object. The slightest disturbance in the air blocks this point. When it becomes unstable and turbulent, it has slightly varying density. In passing through air of varying density, light can be refracted into or out of the path to our eyes. Color changes as the star twinkles, too. When the North Star twinkles, many dimmer stars disappear from view to the naked eye.

The North Star is located on a line established by the two end stars of the dipper at about five times the distance between the dipper stars. Astronomers adjust their brightness scale for stars in terms of magnitudes. The North Star has a magnitude of 2.12. Every five magnitudes corresponds to a factor of one hundred in brightness. By noting which stars twinkle and which tend to disappear from view, you can get some idea of how unstable the air is. The greater the instability, the greater the likelihood for rain.

*When stars flicker on a dark background,*
*Rain or snow follows soon.*

*When dimmer stars disappear, expect rain or snow.*

*When the sky seems very full of stars, expect frost.*

*Excessive twinkling of the stars indicates foul weather.*

✦ ✦ ✦

Much of the time it is difficult to see the dark part of the moon during the new-moon phase even if there are no clouds. This is true because the atmosphere is turbulent. In fair weather, the air is more stable which minimizes turbulence, so we can see more dim objects in the sky. As rainy weather approaches, the air becomes more unstable or more turbulent which tends to obscure dimmer objects in the sky. If we can see the new moon with the old moon in her lap, the air is stable and indicates fair weather.

*If the new moon holds the old moon in her lap, fair weather.*

*If the horns on the new moon are sharper,*
*Fair weather.*

*If the horns on the moon are sharp and pointed,*
*Clear weather, maybe frost;*
*If the points are dull, expect rain.*

*If the moon shows a silver shield,*
*Don't be afraid to reap your field;*
*But if she rises haloed round,*
*Soon you'll walk on flooded ground.*

*Mists in the old moon, rain in the new;*
*Rain in the old moon, mists in the new.*

***If the Indian hunter can hang his powder horn on the crescent of the moon, he should stay home. If he could not hang his powder horn on the moon, he should shoulder his gun and go. The woods will then be quiet.***

This proverb is difficult to analyze because the moon's orbit changes with an 18-year cycle. What's true now will be untrue in nine years, then true again at the end of the next nine years. Also, the moon is sufficiently far from the earth to make all people see the same thing.

The moon's orbit is tilted at about five degrees with respect to the ecliptic plane of the earth's orbit. So, at various times, it is above the earth's orbit and illuminated more from below so the new moon crescent will hold water. Six months later, the new moon will be below the ecliptic plane and tend to be illuminated from above so it will empty water. The periodicity of this is closely related to what astronomers call *seasons of the eclipses of the moon.* If one believes this proverb, his or her observations will vary with an 18-year period. Hence, there are similar but contradictory proverbs.

***If the moon lies on her back,***
***She sucks the wet into her lap.***

***If the crescent of the moon holds water,***
***We will have a dry spell.***

***Tipped moon wet; cupped moon dry.***

*The moon and the weather may change together,*
*But a change in the moon does not change the weather*
*If we had no moon at all, and that may seem strange,*
*We shall have weather that's subject to change.*

✦ ✦ ✦

*Clover leaves show their bottom sides before rain.*

Plants do things to adjust to the sunlight they receive. Chloroplasts can migrate in the leaf cells to get more or less light, giving the plant a different shade of green color depending on the position it is viewed from. Similarly, a leaf can orient itself so it gets optimum sunlight.

Before a cold front arrives with its clouds and rain, the wind tends to be from the southwest. When the wind is in this direction and the plant orients its leaves to get optimum sunlight, the leaves are in an unstable position with respect to wind moving past them. This instability makes the leaves flip over. The unstable plant leaves then herald the approach of a cold front which is likely to bring rain.

*Silver maple leaves turn over before a rain.*

*Cottonwood leaves turn over before a rain.*

*Trees become dark before a storm.*

*Birds stop singing and trees are dark before a storm.*

*Trees are light green when the weather is fair;*
*They turn quite dark when a storm's in the air.*

✦ ✦ ✦

*A veering wind brings fair weather;*
*A backing wind brings foul weather.*

Wind direction is always that *from* which the wind blows. When looking from above, you can observe that the direction is changing in a clockwise direction such as from south to west. We say it is *veering with time.* If the wind is shifting in a counterclockwise direction such as from south to east, we say the wind is *backing with time.*

We must also remember that wind motions are clockwise about a high-pressure center and are counterclockwise about a low-pressure center. It also is true that low-pressure centers tend to move from the southwest toward the northeast. If the wind is veering, a low-pressure center is most likely passing us with a path to our north, while if the wind is backing, the low-pressure center is most likely passing with a path to our south.

*The warmth of the south wind is enervating;*
*The cold of the north wind is bracing;*
*The chill of the east wind brings aches and pains;*
*The prevailing west winds and moderate temps*
*Impart the dominant qualities that are possessed*
*By the people of the temperate zones.*

*When the sun sets bright and clear,*
*An easterly wind you need not fear.*

If you have watched television reports of satellite pictures, you may remember that a low-pressure weather system often appears as a large "comma" with lots of precipitation in the head of the comma. A backing wind will tend to place us in the head of the comma.

✦ ✦ ✦

***If the clouds move against the wind, rain will follow.***

When clouds move against the wind, we have what is called *wind shear.* Wind shear is strong along a passing cold front, so very likely a cold front is approaching. Clouds are moved by upper winds that will be in a different direction from the lower winds.

We say that winds *veer with altitude* if they are changing in a clockwise turning as one goes upward, such as an east wind at the ground and a south wind above. We say that winds are *backing with altitude* if they are turning counter-clockwise as flow goes upward, such as shifting from east at the ground to north above. Meteorologists know well that when winds veer with altitude, we will have a warming trend. When winds back with altitude, we will have a cooling trend.

By watching the clouds we can tell a lot about tomorrow's temperature. To remember this rule we have "backing" with "cooling" at the beginning of the alphabet and "veering" with "warming" at the end of the alphabet.

***There's little use in praying for rain***
***If the wind is in the north.***

***When the sun sets in a bank [of clouds],***
***A westerly wind we shall not want.***

***When the wind's in the west,***
***The weather's always best.***

***A northern air brings weather fair.***

✦ ✦ ✦

*If the wind's in the north,*
*The skillful fisherman goes not forth;*
*If the wind's in the east,*
*It's good for neither man or beast;*
*If the wind's in the south,*
*It blows the fly in the fish's mouth;*
*When the wind is in the west,*
*There it is the very best.*

Isaak Walton

Isaak Walton was a great naturalist who probably knew a lot about fishing. When the wind is in the west, the weather is, in general, the best. From experience, I say that the fishing is not necessarily the best. It seems that fishing is best when pressure is falling, which allows food to be released from lake and river bottoms so fish start feeding. In the case for walleye-pike fishing, one usually does best with not too large, but choppy waves on the lake. This often occurs with west winds. I think Walton was primarily a trout fisherman, but it may be that he also fished on English lakes. It seems each and every fisherman has his own rules about fishing. When he learns a good rule, he is not likely to tell anyone about it.

*When smoke goes west,*
*Good weather is past;*
*When smoke goes east,*
*Good weather comes niest [next].*

Scottish

*When the wind's in the south,*
*There's rain in its mouth.*

*Southerly winds with showers of rain,*
*Will bring the wind from the west again.*

✦ ✦ ✦

When ants find food, they release a *pheromone* as they move back to the ant colony. Other ants then smell the pheromone along the trail to the source of food. The pheromone is a very volatile substance, meaning it won't stay in place very long. However, when the humidity is high, water molecules in the air can attach to the complex pheromone molecules and make them less volatile so they will stay in place longer and allow many ants to follow the trail. In dry air, the pheromone is so volatile that it vanishes before the ants can form a trail so the ants simply wander more or less at random.

***Expect stormy weather when ants travel in a straight line; when they scatter all over, the weather is fine.***

Experimenters have herded an ant that has found food into complex spiral paths. They find that all the ants will then smell their way along the spiral path instead of a straight line. Similarly, some processional caterpillars travel in lines. If the lead caterpillar is herded to the tail of the line, the caterpillars will move in a circle for days.

***If ants carry eggs to high ground,***
***Expect rain.***

***Ants building sand cones around holes,***
***Expect rain.***

***Ants are very busy before a rain.***

✦ ✦ ✦

***Flies bite greedily before a rain.***

Most of us have experienced this, but it is a little difficult to know why it is true. One of many things different about insects is their respiratory systems. Instead of having lungs to breath in air for oxygen, tracheal tubes extend from the outside to all interior parts of their bodies. When one has a very small capillary tube such as a tracheal tube, water molecules from the air can enter the tube and condense to fill it with liquid at vapor pressures far below saturation pressures in the air. This liquid in the tubes cuts off air circulation. Why this makes the fly hungry for our blood, I don't know. Perhaps he wants blood for the oxygen in it.

An increase in humidity would then make the fly hungry and indicate to us that more moist air was moving in to give us rain. Insects have blood, but it is used to carry food only, not food and oxygen as does our blood system. Also, flying aids breathing in insects. As the small-capillary breathing tubes become plugged with condensing water vapor when the humidity rises, insects may simply become more mobile and hence encounter us more often.

*If flies come in great swarms,*
*Rain follows soon.*

*When eager bites the thirsty flea,*
*Clouds of rain you're sure to see.*

*Fireflies are out before a rain.*

✦ ✦ ✦

***If fog clears from the top down, expect some rain.***

Usually fog forms during high-pressure times when the air above is dry and there can be lots of cooling through energy loss by radiation. The earth, being a better radiator than air, gets colder than the air. Temperature, therefore, increases with altitude to form a *temperature inversion.*

When fog clears from the top down, it means the fog is very dense. In turn, this implies lots of water vapor in the air. Eventually the fog will burn off, but the air will acquire a large relative humidity in doing so. We now have somewhat contradictory conditions—the pressure is likely high, a fair-weather condition; and relative humidity is high, a condition that leads to rain.

If the sun continues to shine, this moist air near the ground can start to form large, warm bubbles that will rise and eventually cool to condensation. The heat liberated in condensation can make the air more unstable and clouds can build to a shower. In contrast to frontal showers, these are called *land-mass showers.*

***If fog clears from the bottom up,***
***The weather will be fair.***

***An old moon in a mist***
***Is worth gold in a kist [chest];***
***But a new moon's mist***
***Will never lack thirst.***
Scottish

✦ ✦ ✦

***Lamp wicks crackle before a rain.***

It seems as if kerosene in lamp wicks can take moisture from the air. Oil is a *non-polar* molecule and water is a *polar* molecule. This is why they don't mix. Any absorbed water can collect in pockets. When the flame burns, the heat changes the water to steam and makes the flame sputter. Oily floors also take up water in layers and become very slippery.

Campfires don't burn well when relative humidity is high. Two things can make this true: one is that capillaries in the wood have taken in some moisture; the other is that water vapor in the air decreases the amount of oxygen in the air. Part of the air is water vapor, and high humidity is necessary for rain.

***Candles burn dimmer before a rain.***

***Campfires sputter and spit before a rain.***

***Campfires burn brighter in fair weather.***

***Coals covered with thick white ashes***
***Indicate snow in winter and rain in summer.***

***A fire hard to kindle indicates bad weather.***

✦ ✦ ✦

***The farther the sight***
***The closer the rain.***

There are many complicated phenomena involving both sight and sound. When rain is close, the air becomes quite humid or there are more water molecules in the air. Water molecules are lighter than air molecules. Distant vision is obscured because of haze that comes from aerosols in the air. If aerosol particles are very small, they will not scatter light. When their size reaches a fraction of the wavelength of light, they can scatter the light, preferably blue light. So haze tends to be blue. As humidity rises, water molecules collect on aerosol particles and make them grow so at first the air becomes more hazy.

Before rain, the air becomes unstable and these particles drift upward toward the cloud base to clear the air. Lighter water molecules increase the velocity of sound through air. As air rises, the percentage of water vapor in the air remains constant, so refraction of sound depends only on the changing temperature. The cooler temperature will make sound refract upward and then back down from cloud bases, thus reflecting sound waves from distant sources back to ground level so we hear more distant noises.

***Sound travelling far and wide,***
***A stormy day will betide.***

***If waterfalls roar loudly, bad weather is coming.***

***Distant objects appear to be closer before rain.***

***Sailors say that mirages indicate a coming storm and lower their sails.***

✦ ✦ ✦

***Mists, if they rise on low ground and soon vanish, fair weather.***

Fog forms when there's little water vapor in the upper air. Heat from ground level can then radiate readily to outer space and cool the ground to dew-point temperature. First, there will be dew. As the air next to the ground cools, fog forms. If the fog is shallow, there's not much vapor in the air.

If a fog bank becomes thick there is more moisture—maybe enough to lead to rain. When mists are shallow, bright sun can burn them off from above and below as air temperature rises above the dew-point temperature.

Note: Practically all water vapor in the air is usually in the lower two miles of the atmosphere.

*If mists rise to the hill tops,*
*Rain in a day or two.*

*When the mist comes from the hill,*
*Then good weather it doth will;*
*When mist comes from the sea,*
*Then good weather it will be.*

*In the decay of the moon,*
*A cloudy morning bodes a fair afternoon.*

*Rain before seven,*
*Clear before eleven.*

✦ ✦ ✦

***Dogs hunt better before a rain.***

Dogs do most of their hunting by sense of smell. Large, aromatic molecules given off by the quarry are picked up by the dog who can follow a concentration gradient of these molecules. When the air is moist or relative humidity is high, water molecules in the air tend to attach to the large molecules. With the layer of water molecules on the large molecule, they can attach more readily to sense organs in the animal's nose and will enhance the sense of smell. (Most animals have a more acute sense of smell than we do.) A higher humidity makes rain more likely.

*If dogs and horses sniff the air,*
*A summer shower will soon be there.*

*If cows and sheep sniff the air,*
*A sign of rain.*

*If horses stretch out their necks and sniff,*
*Rain will ensue.*

*If asses hang their ears downward and forward,*
*and rub against the wall, rain is approaching.*

✦ ✦ ✦

*Cats and dogs eat grass before rain.*

Cats and dogs eat grass because they feel distress in their stomachs and need to vomit. This distress is most likely due to a barometric-pressure change. When pressure is high in fair weather, an animal's body must absorb gas to keep it from collapsing. When pressure falls, these absorbed gases must again be released to avoid exploding.

Gas can be absorbed in a body liquid molecule by molecule, but when the gas is released, it is always nucleated into bubbles. The animal is trying to relieve itself of the gas bubbles. Often there is also irritation of the skin in the form of itching and resulting scratching when the gas bubbles are released. The falling barometer indicates a possibility of rain.

*Dogs rolling on their back expect rain.*

*Horses rolling over in the pasture, expect rain.*

*Chickens puddling in the dust predict rain.*

*Young asses rolling and rubbing their back on the ground*
*indicates heavy showers.*

✦ ✦ ✦

# 6 Long-range Proverbs

***The winter will be severe if the roughed grouse have heavier than usual fringes on their toes.***

Weather lore we've looked at up to this point can be classified as *short range* because it deals with weather predictions covering a few days. This is the lifetime of a weather system as it forms and moves across North America. Short-range proverbs tell us things about many of the variables that meteorologists must know to make their predictions for the next few days. Most short-range proverbs tell us something we want to know about how meteorological variables change.

If we ask more long-range questions—Why do weather systems form?, When and where will they form?, How long will they last?, In which direction will they travel?—even the long-range technologies of the Weather Service do not work very well.

Experts know weather systems form to distribute energy from the sun from equator to pole and to form a giant heat engine that drives atmospheric motions. They know that in transporting this energy there is a problem of correctly changing the angular momentum of the atmosphere to yield the general swirling motions generated by a weather system. But they cannot predict where these systems will form, how large they will be, how they will move, and how long they will last. At present, we have insufficient knowledge about the physics of the atmosphere to make such predictions.

Every once in a while there is some evidence that there can be success in making long-range predictions. There were examples of forecasting in the long-range sense before there was a science of meteorology. I describe a few of these examples and then simply list many of the long-range proverbs. We should watch these proverbs for a clue to how plants and animals may have developed some long-range sensing of what may happen in our atmosphere during the next year.

Let's start with a few success stories: Aristotle described how Thales of Melitus became very rich. Thales knew astronomy and many weather proverbs of the time. After several years of rather average olive-crop yields, he predicted that the next summer would lead to a very large yield of olives. He immediately started to buy up all the olive-oil presses in the land. The next year brought a very bountiful crop of olives and he completely controlled the olive-oil market.

Virgil wrote a book entitled *Georgics* in which he gave a long list of rules for successful farming in Italy—all given in the form of proverbs.

Our prevailing winds over the globe are easterly winds from the equator to about 20 degrees of latitude either north or south separated at the equator by the *doldrums* where there often is little wind. From 20- to about 25-degrees latitude are the *horse latitudes* where again there is little wind. North of this we have prevailing westerlies to around 55 degrees north or south. Most weather systems form around 55 degrees north or south. Above these latitudes and closer to the pole, we have prevailing easterlies again. Meteorologists understand this well in terms of *Hadley Cell circulations* combined with *Coriolis forces.*

***Between the tropics the winds and currents tend westward.***

***In middle latitudes, winds and currents tend eastward.***

***In high latitudes, winds and currents tend from the poles toward the equator.***

These describe general circulation in the Hadley cells. Air rises over the equator and descends at about 25 degrees north and south latitude so that air is returning to the equator. The Coriolis force deflects wind toward the right in the northern hemisphere and to the left in the southern hemisphere to give easterly trade winds. A second cell gives rising air at about 50 degrees north and south which descends again at 25 degrees. Deflection of these winds by the Coriolis force gives us prevailing westerlies. A third cell over the poles tend to give southward flow with deflections toward the west, resulting in polar easterlies. For some unknown reason, easterlies die away over the Pacific and lead to a phenomenon called an *El Niño.*

Columbus did not know this meteorology, but he had sailed enough to know that to sail westward you should take advantage of the prevailing easterlies and set a course just north of the equator. While sailing eastward, you should take advantage of prevailing westerlies and set a course north of the horse latitudes. With this knowledge, he made several trips across the Atlantic Ocean from Europe to America.

The Pilgrims, on the other hand, hired rather stupid sea captains. They tried to sail from Europe to America against the prevailing westerlies. These trips took so long that many passengers died on the way.

Many sailing ships became stalled in the horse latitudes where there is little wind. They were explorers and often had horses on board. When they became stalled, they would run out of water and, for purposes of conservation, threw the horses overboard. Hence the name *horse latitudes.*

Prevailing winds create ocean currents. North and south of the equator in the Atlantic, currents are from east to west, driven by the prevailing easterlies. When these currents hit the bulge of Brazil in South America, they turn north and south, creating the Gulf Stream in the Northern Hemisphere. Smart sea captains knew they could cross the doldrums or horse latitudes by means of these currents. They knew how to do all of this, but they didn't know why.

World-wide sailing from the early 15th century through the early part of the 19th century required about 124 days for English navigators to sail from England to Australia. Most of this time was spent bucking prevailing easterlies—as they sailed in a southeastern direction along the west coast of Africa—or getting stuck in the equatorial doldrums.

Captain Maurey of the U.S. Navy pointed out that they should sail westward with prevailing easterlies north of the equator to South America. Then let ocean currents carry the ship across the doldrums and continue southwesterly in the southern easterlies until they crossed the southern horse latitudes and got into the prevailing westerlies in the southern hemisphere. These westerlies would rapidly carry the ship to Australia, so the total time for the trip was reduced to 93 days. Again, they did not know the meteorological reasons for their procedure, but their observations were reliable.

Today, the planet earth is often plagued with an *El Niño* in which the eastern Pacific Ocean, at equatorial latitudes, becomes very warm. For some reason, the prevailing easterlies across the Pacific tend to vanish. Along with this, ocean currents in the Pacific tend to vanish. There is, then, little upwelling of cold water along the western coast of South America and fishing becomes very poor. Elaborate measurements of Pacific Ocean water show when this warming trend is taking place.

Something that accompanies this phenomenon is that the bird population vanishes on Christmas Island, an equatorial island in the middle of the Pacific. Somehow the birds are aware of what is happening.

These long-range proverbs come from many countries of the world and all say basically the same thing. They were developed when communications between countries were very limited, so the proverbs were developed independently. It seems that different people making observations in different lcoations all came to the same conclusions.

At present, we do not understand all the physical reasons as to why there may or may not be truth in them. I show why a few of them may have some truth according to what we now know about weather-system formation. Because short-range proverbs were formed without understanding as to why they were true and yet some proved to be accurate, it seems sensible to list the long-range proverbs and look for clues as to what they may tell us about the weather.

✦ ✦ ✦

***If Candlemas be fair and clear,***
***There'll be two winters in the year.***

Candlemas Day, known by many as *Groundhog Day,* is about one week after the coldest day of winter. Maybe there's a hint of truth in this long-range forecast. High- and low-pressure systems are moving through the region as always.

If Candlemas Day is clear, we are experiencing a high-pressure system that may last for a week before a low-pressure system moves in. This is followed by yet another high that also may be quite cold. By the time this is over, the days are getting much longer, the sun is higher in the sky and there is general warming. But we may have experienced three more weeks of winter weather.

On the other hand, if it is cloudy on Candlemas Day, we are experiencing a low to be followed by a high and another low. The temperature is usually considerably higher during low pressure so that even though there is a high sandwiched in between two lows, we tend to think of winter as having its *back broken.* Don't pack away your fur coat if it's clear.

***On the eve of Candlemas Day,***
***The winter gets stronger or passes away.***
***If the groundhog sees his shadow on February second,***
***There will be six more weeks of winter.***

***Candlemas Day, Candlemas Day,***
***Half your wood, half your hay.***

***If Marie's purifying daie***
***Be bright and clear with sunny naie,***
***Then frost and cold shall be much more,***
***After the feast than was before.***

***If it neither rains nor snows on Candlemas Day,***
***Saddle your horse, buy him some hay.***

***When on the purification, the sun has shined,***
***The greater part of winter comes behind.***

***If Candlemas be fair and bright,***
***Winter will have another flight;***
***But if Candlemas brings clouds and rain,***
***Winter is gone and won't come again.***

***Good weather on this day indicates a long***
***continuation of winter and a bad crop;***
***On the contrary, if foul, it is a good omen.***

*If Candlemas be mild and gay,*
*Go saddle your horses and buy them hay;*
*But if Candlemas be stormy and black,*
*It carries the winter away on its back.*
*If Candlemas be dry and fair,*
*The half of winter's to come and mair;*
*If Candlemas be wet and foul,*
*The half of winter was gone at Yule.*

*After Candlemas Day the frost will be more keen,*
*If the sun shines bright, than it has been.*

*On Candlemas Day, the bear, badger, and woodchuck*
*Come out to see their shadow at noon;*
*If they do not see it, they remain out,*
*But if they see it, they return to their holes*
*For six more weeks of winter.*

*If the groundhog is sunning himself on 2nd February,*
*He will return to his winter quarters for*
*Six more weeks of winter.*

*The badger peeps out of his hole on Candlemas Day,*
*And when he finds it's snowing, walks abroad;*
*But if he sees the sun shining,*
*He draws back into the hole.*

*At the day of Candlemas,*
*Cold in air, and snow on grass;*
*If the sun then entice the bear from his den,*
*He turns around thrice and goes back again.*

*When it rains on Candlemas,*
*the cold is over.*

*If the lanes are full of snow on Candlemas,*
*So the bins will be full of corn in autumn.*

✦ ✦ ✦

*The ash before the oak*
*Choke, choke, choke,*
*The oak before the ash*
*Splash, splash, splash*

"Before" refers to "budding before." "Choke" refers to dry weather, and the "splash" refers to wet weather. The proverb says that if oak trees bud before ash, we will have a wet season. But if ash trees bud before oak, we'll have a dry season.

The budding of trees is determined by moisture in the soil as well as temperature. If we have had a dry fall and winter, there is little moisture in top soil, but the deeper soil may be moist from rain the previous summer. If we have had a wet fall and winter, all the soil is moist. Oak has a deep tap root that can reach to deeper moisture; ash has a more shallow root system that is more dependent on the history of moisture added to the soil. If we have had a wet fall and winter, ash is more ready to bud and will beat the oak. But if the previous seasons have been dry, oak is more ready to bud than ash. The proverb is a statement of nature trying to maintain precipitation averages.

*When the ash is out before the oak,*
*Then we may expect a choke;*
*When the oak is out before the ash,*
*Then we may expect a splash.*

*When buds the oak before the ash,*
*You'll only have a summer splash.*

*If the oak is out before the ash,*
*'Twill be a summer of wet and splash;*
*But if the ash before the oak,*
*'Twill be a summer of fire and smoke.*

✦ ✦ ✦

*June warm and damp*
*Does not make the farmer*
*poor.*

*March flowers*
*Bring summer showers.*

*No killing frost after*
*martins are seen.*

*When April blows his horn,*
*It's good for both the hay*
*and corn.*

*Unusually thick nutshells*
*predict a severe winter.*

*The first thunder of the year*
*awakes all the frogs and snakes.*

*Thunder in March, a fruitful harvest.*
*A dry spring, rainy summer.*

*If we do not get an Indian Summer in*
*October or November,*
*We will get it during the winter.*

*A moist autumn with a mild winter*
*is followed by a cold dry spring.*

*A moist and cool summer portends a hard winter.*

*Winter thunder bodes summer hunger.*

*A swarm of bees in May*
*is worth a load of hay.*

*If the spring be cold and wet,*
*the autumn will be cold and dry.*

*When leaves fall early, fall and winter will be mild;*
*when leaves fall late, winter will be severe.*

*Who doffs his coat on a winter day*
*gladly puts it on in the month of May.*

*Better late spring and bear*
*than early blossom and blast.*

*An early winter,*
*a surly summer.*

*Early thunder,*
*early spring.*

*If bears lay up food*
*in the fall, it indicates*
*a cold winter.*

*Easter in snow,*
*Christmas in mud;*
*Christmas in snow,*
*Easter in mud.*

*As high as the weeds grow,*
*so will be the bank of snow.*

*May damp and cool*
*fills the barns with wine vats.*

*There never can be too much rain before midsummer.*

*A late spring is good for corn but not cattle.*

*When squirrels lay in a large supply of nuts, there will be a severe winter.*

*If birds migrate early, we'll have a severe winter.*

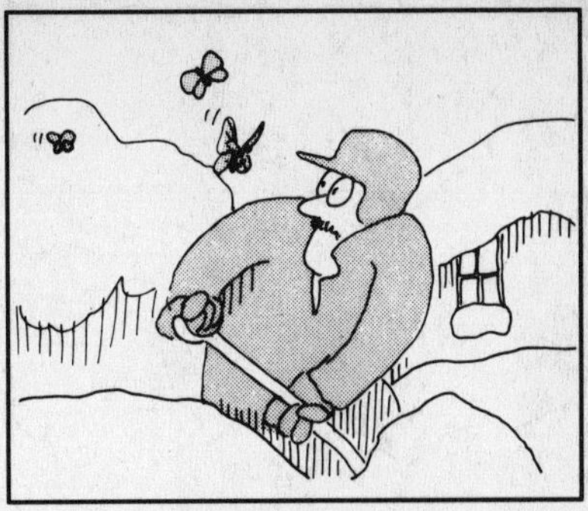

*The early appearance of butterflies indicates fine weather.*

*If the November goose bone be thick, so will the winter weather be.*

*If the breastbone of the Thanksgiving goose is red or has many red spots, expect a cold and stormy winter; but, if only a few spots are visible, we will have a mild winter.*

*If March comes in like a lion, it goes out like a lamb; if March comes in like a lamb, it goes out like a lion.*

*If crows fly south, we'll have a severe winter.*

*Wet May, dry July.*

*A cold and wet June spoils the rest of the year.*

*As August, so next February.*

*March winds and April showers bring May flowers.*

*If squirrels are scarce in autumn, it indicates a cold winter.*

*If trees hang onto their leaves, the coming winter will be cold.*

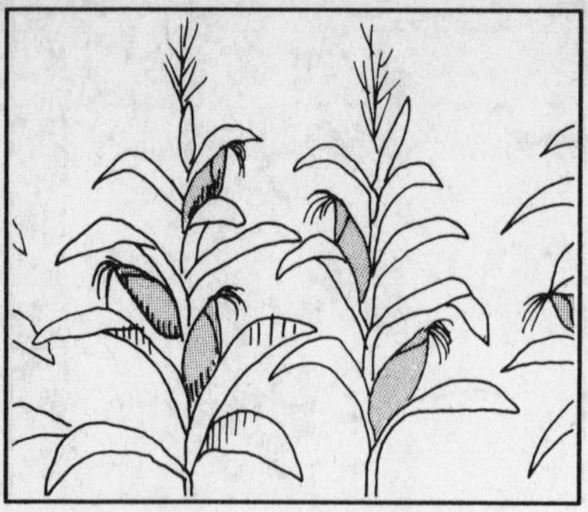

*Calm weather in June*
*sets corn in tune.*

*Thick and tight cornhusks*
*predict a hard winter.*

*February rain is good only*
*to fill ditches.*

*A great store of nuts,*
*a good corn year.*

*A snow year is a rich year.*

*Onion skins very thin*
*Mild winter coming in;*
*Onion skins thick and tough*
*Coming winter cold and rough.*

*Plant your beans when the moon is light,*
*You will find that this is right;*
*Plant potatoes when the moon is dark,*
*And to this line you'll always hark;*
*But if you vary from this rule,*
*You will find you are a fool;*
*Follow this rule to the end,*
*And you'll have lots of dough to spend.*

*If the cock moult before the hen,*
*We'll have winter thick and thin;*
*But if the hen moult before the cock,*
*We'll have winter hard as rock.*

*A green winter makes a fat church yard.*

*If you take an "oak apple" from the oak tree and open the same, you will find a little worm therein, which, if it doth creep, it betokens a scarcity of corn.*

*Don't plant your corn until the oak leaf is as big as a squirrel's ear.*

*A wet September, drought next year.*

*A windy March and a showery April makes a beautiful May.*

*Much hair on animals predicts a cold winter.*

*The severity of the winter can be determined by how far down the feathers grow on a partridge's leg.*

*In early and long winters the beaver cuts his supply of wood earlier than in mild and late winters.*

*A January spring is worth naething.*

*If moles have only a few worm basins, we will have a mild winter.*

*Chipmunks are seen as late as December before a mild winter.*

*A month that comes in good will go out bad.*

*Always expect a thaw in January.*

*If there is no snow in January, there will be more in March and April.*

*A cold April the barn will fill.*

*When squirrels store nuts*
*high in trees, we will have*
*a winter of deep snow.*

*If the walnut tree has*
*plenty of blossoms,*
*a sign of a fruitful year.*

*When there's lots of snow,*
*a fruitful crop will often*
*grow.*

*When the oak puts on its*
*gosling gray,*
*it's time to sow barley*
*night and day.*

*The whiteness of a goose bone indicates*
*the amount of snow.*

*Fall bugs begin to chirp six weeks before a frost.*

*He who sows his oats in May will get little that way.*

*If October and November be snow and frost,*
*then January and February are likely*
*to be open and mild.*

*Early insects, early spring, good crops.*

*The wider the band on the woolly caterpillar,*
*the milder the winter.*

*A warm January, a cold May.*

*January warm, the Lord have mercy;*
*A favorable January, a good year.*

*There is always one fine week in February.*

*If February gives much snow,*
*a finer summer it doth show.*

*December cold with snow,
good for rye.*

*If it rains in August, it
rains honey and wine.*

*As September, so the
coming March.*

*Dry August and warm does
harvest no harm.*

*If October brings heavy
frosts and cold winds,
January and February will
be mild.*

*When skunks are real fat,
we'll have a long
winter coming.*

*As the weather in October,
so will be the next March.*

*Sow peas and beans in the wane of the moon;
Who soweth them sooner, soweth too soon.*

*Plant your beans when the moon is light;
Plant potatoes when the moon is dark.*

*Root crops are planted in the dark of the moon.*

*Leaf and grain crops are planted
in the light of the moon.*

*Much rain in October,
much wind in December.*

*If the cuckoo sings when the hedge is green,
keep thy horse and sell thy corn.*

*If mud wasps build their nests in sheltered areas,
expect a harsh winter.*

*If a cow's droppings are frozen in September,
they will thaw in October.*

*April snow is as good as lambs's manure.*

*If the pig wallows in a puddle before the first of May,*
*the summer will be cold.*

*When walnut trees bear bountifully,*
*we will have a warm winter.*

*The first snow comes six weeks after the*
*last thunderstorm in September.*

*Moist April, clear June.*

*As high as the hornet's nest*
*so will be the snow next winter.*

*After the robin comes in spring,*
*he'll get snow on his back three times before it stops.*

*When muskrats build larger houses in deeper water,*
*it's going to be a cold winter.*

# 7 Days & Dates

*If it rains on Easter Sunday,*
*it will rain for seven Sundays in a row.*

We tend to remember better what happens on holidays than on other days. This is simply because we do something special which reinforces all of the events on these special days. Often the special event is an outdoor activity that is subject to weather conditions. If it rained at every 4th of July picnic for the last four years, we would likely conclude that it always rains on the 4th of July. But if it rained every 4th of June, we probably wouldn't remember it. There have been many feast days and holidays through history. The weather on these days is remembered and shows up as proverbs.

There probably is little truth in these proverbs. If weather were associated with special days, there would be an annual periodicity in weather patterns. Meteorologists have subjected weather data to many studies in order to find periodicities that may help in forecasting, but usually end up finding none.

Still there are interesting things about average values of weather parameters such as temperatures during the seasons, how we should divide our seasons, and how attempts have been made to make our calendar reflect the nature of our seasonal climate rather than some Roman God or King's birthday.

We start this chapter by looking at a few of these things, then proceed to list many of the proverbs that have been associated with special days.

✦ ✦ ✦

| | |
|---|---|
| **Winter:** | St. Catherine's Day (Nov. 23) to 21st of February |
| **Spring:** | St. Peter's Day (Feb. 22) to 24th of May |
| **Summer:** | St. Urban's Day (May 25) to 23rd of August |
| **Fall**: | St. Bartholomew's (Aug. 24) to 22nd of November |

Our present division of the calendar comes to us from astronomers, which doesn't make too much sense as far as weather is concerned. Prior to our present division, the division of seasons was as described above. The old division made the seasons much more symmetrical about the coldest day of the year toward the end of January and the warmest day of the year toward the end of July. So many times we have big snow storms in early December and people remark, "and it isn't even winter yet!" Similarly, we have very hot weather early in June and people will remark, "and it isn't even summer yet." The old division of seasons makes much more sense as far as the weather and climate is concerned.

✦ ✦ ✦

*Hoppy, croppy, poppy;*
*Wheezy, sneezy, breezy;*
*Nippy, drippy, slippy;*
*Showery, flowery, bowery.*

The French, after their revolution, changed everything—even the calendar. The new calendar started September 1, 1792. The months were named *Vendemiare* (Vintage), *Brumaire* (Mist), *Frimaire* (Frost), *Nivose* (Snow), *Phiviose* (Rain), *Ventose* (Wind), *Germinal* (Budding), *Floreal* (Flowering), *Prairial* (Meadows), *Messidore* (Harvest), *Thermidore* (Warmth), *Fructidore* (Fruit). The seasons started with *fall* on the first of Vendimaire, *winter* on the first of Nivose, *spring* on the first of Germinal, and *summer* on the first of Messidore. This made for a much better division of the seasons as far as climate is concerned. As usual, there were folks who did not think much of a new idea, so they referred to the months as in the above poem.

✦ ✦ ✦

This proverb is a bit of Saxon lore. They tell us that if the New Year falls on Sunday, we will have a good winter, a windy spring and a dry summer; on Monday, we will have a severe winter, a good spring and a windy summer; on Tuesday, a dreary winter, a windy spring and a rainy summer; on Wednesday, a hard winter, a bad spring and a good summer; on Thursday, a good winter, a windy spring and a good summer; on Friday, a variable winter, a good spring and a good summer; and on Saturday, a snowy winter

*The trend for the year's weather is determined by the day on which New Year falls.*

with a rainy spring. From this, it looks as if Friday is the best day for New Year's Day.

✦ ✦ ✦

***The twelve days of Christmas determine the weather for each month of the year.***

This English weather lore tells us that we should note the weather on each of the twelve days which will then correspond to the weather for the corresponding months. The first day of Christmas is December 25; the second day is known as Boxing Day. In wealthier families, the Christmas feast is on December 25 and the servants have to work to feed the family. The servants then have the next day off and receive their *boxes.* The twelfth day of Christmas is January 5, the Eve of Epiphany, and decorations are removed on January 6.

✦ ✦ ✦

This poem gives a remarkably good way to remember the average temperatures during each month of the year.

***Round the whole of the year***
***Forty-three stands for the mean,***
***For July and August,***
***Add two times thirteen;***
***Put fourteen on for May***
***Twenty-three for June,***
***And remember to allow***
***Two times eight degrees***
***For the heat of September.***

***October and April improve***
***Four degrees on the year,***
***But the list of warm months***
***Really ends here.***

*Cold January's down*
*By thirty-two degrees*
*December and February,*
*Twenty-seven for these.*

*Make it two times seven*
*For March and November;*
*Now isn't this easy*
*For you to remember.*

✦ ✦ ✦

**January 14: St. Hilary's Day**

*The coldest day of the year.*
(In Minnesota, the coldest day is January 20.)

**January 17: St. Sulpicius' Day**

*Frost on St. Sulpicius day augurs well for the spring.*

**January 22: St. Vincent's Day**

*Sunny on St. Vincent's,*
*we shall have more wine than water.*

*If St Vincent's has sunshine,*
*we get much rye and wine.*

**January 25: St. Paul's Day**

*If St. Paul's be fair and clear,*
*it doth betide a happy year;*
*if by chance it then should rain,*
*it will make dear all kinds of grain.*

**February 1: St. Bridget's Day**

*Bridget's feast day white, every ditch full.*

**February 2: Candlemas Day**

*More snow and ice if the sun shines*
*on Candlemas Day.*

**February 6: St. Dorothea's Day**

*St. Dorothea gives the most snow.*

**February 12: St. Eulalie's Day**

*Sun smiles; good for apples and cider.*

**February 14: St. Valentine's Day**

*Winter breaks her back.*

**February 22: St. Peter's Day**

*The night of St. Peter's shows the*
*weather for the next forty days.*

**February 24: St. Mathias Day**

*If it freezes on St. Mathias,*
*it freezes for a month together.*

**February 28: St. Romanus Day**

*Romanus bright and clear indicates a goodly year.*

**March 1: St. David's Day**
**March 2: St. Chad's Day**
**March 3: St. Winnelow's Day**

*First comes David,*
*Then comes Chad*
*And then comes Winnelow*
*As though he were mad.*

**March 17: St. Patrick's Day**

*The warm side of a stone turns up,*
*and the broadback goose begins to lay.*

**March 19: St. Joseph's Day**

*St. Joseph's Day clear*
*so follows a fertile year.*

**March 21: St. Benoit's Day**

*If it rains on St. Benoit's,*
*it will rain for forty days after.*

**March 25: St. Mary's Day**

*St. Mary's bright and clear,*
*fertile is said to be the year.*

**April 5: St. Vincent's Day**

*If St. Vincent's be fair,*
*there will be more water than wine.*

**Palm Sunday**

*A bad Palm Sunday will*
*lead to poor crops.*

**Easter**

*If it rains on Easter Sunday,*
*it will rain for seven*
*Sundays in a row.*

**April 1: All Fools' Day**

*If it thunders on All Fools' Day, it will bring good crops of corn and hay.*

**April 23: St. George's**

*If it rains on St. George's, he eats all the cherries.*

**May 8: St. Phillip's, St. James' Day**

*If it rains on St. Phillip's and St. James',*
*a fertile year may be expected.*

**May 11: St. Mamertus' Day**
**May 12: St. Pancras' Day**
**May 13: St. Servatius' Day**

*St. Mamertus, St. Pancras, and St. Servatius*
*do not pass without a frost.*
*He who shears his sheep before St. Servatius*
*loves his wool more than his sheep.*

**May 19: St. Dunstan's Day**

*The devil may bring a frost to blast the apple crop—*
*a rival for beer.*

**May 25: St. Urban's Day**

*This day inaugurates summer.*

**May 26: St. Phillip's Day**

*When it rains on St. Phillip's,*
*the poor will need no help from the rich.*

**June 8: St. Medard's Day**

*If it rains on St. Medard's,*
*it will rain for forty days.*

**June 11: St. Barnabus' Day**

*On St. Barnabus' Day, the sun comes to stay.*

**June 15: St. Vitus' Day**

*If St. Vitus day be rainy,*
*it will rain for thirty more days.*

**June 19: St. Protase's Day**

*If it rains on St. Protase,*
*it will rain for forty more days.*

**June 24: St. John's Day**

*Rain on St. John's Day, expect a wet harvest.*

**June 29: St. Peter's Day**

*If it rains on St. Peter's and St. Paul's,*
*the bakers will have to carry*
*double flour and single water.*
*If it is clear, they will have to carry*
*single flour and double water.*

**July 4: St. Martin Bullion's Day**

*If Bullion's day be dry,*
*there will be a good harvest.*

**July 14: St. Processus' Day, St. Martinian's Day**

*If it rains on the feast of Processus and Martinian,*
*it suffocates the corn.*

**July 15: St. Swithin's Day, St. Gallo's Day**

*If St. Swithin weeps, it will rain for forty days.*

*The weather on St. Gallo's will prevail for forty days.*

**July 19: St. Vincent's Day**

*Rains cease and winds come.*

**July 20: St. Margaret's Day**

*So much rain falls that they call it Margaret's flood.*

**July 22: St. Mary Magdalene's Day**

*St. Mary washes her handkerchief.*

**July 26: St. Anne's Day**

*If it rains on this day,*
*it will rain for one month and one week.*

**July 27: St. Godelieve's Day**

*If it rains on this day,*
*it will rain for forty days after.*

**August 10: St. Laurence's Day**

*If the weather is fair,*
*fine autumn and good wine.*

**August 15: St. Mary's Day**

*On St. Mary's sunshine, much good wine.*

**August 24: St. Bartholomew's Day**

*If this day is misty with frost in the morning,*
*cold weather will come soon with a cold winter.*

**September 21: St. Matthew's Day**

*St. Matthew's rain fattens pigs and goats.*

**September 29: Michaelmas Day**

*So many days old the moon on Michaelmas,*
*so many floods after.*

**October 18: St. Luke's Day**

*Fine dry weather on Luke's,*
*little summer.*

**October 28: St. Simon's Day, St. Jude's Day**

*A rainy day.*

*On St. Jude's Day the oxen may play;*
*November take flail let no more ships sail.*

**November 1: All Saints' Day**

*Some weather that is warm is called "All Saints' Rest" (Indian Summer).*

**November 11: St. Martin's Day**

*If the leaves do not fall before St. Martin's, expect a cold winter.*

**November 23: St. Catherine's Day, St. Clement's Day**

*This day inaugurates winter.*

**November 26: St. Vincent's Day**

*On St. Vincent's, winter waxes or wanes.*

**December 21: St. Thomas' Day**

*Look at the weather cock on St. Thomas at twelve o'clock and see which way the wind is for there it will stick for the next lunar quarter.*

**December 25: Christmas**

*A green Christmas, a fat churchyard.*

*If windy on Christmas, a fruitful harvest.*

**December 26: St. Stephen's Day**

*St. Stephen's Day windy, bad for next year's grapes.*

# 8 Trouble With Some Proverbs

***Red skies in the morning, sailor's warning;***
***Red skies at night, sailor's delight.***

It's difficult to find much truth contained in some proverbs. I have often thought that some *one* proverb can't have truth; yet, on further consideration, I find truth in the saying. The best example of this is the proverb above.

If the red is in the *eastern* sky in the morning and in the *western* sky in the evening, the proverb is likely to be wrong. If the red is in the *western* sky in the morning and in the *eastern* sky at night, the proverb may be true. We are not sure what the writer had in mind when the proverb was written.

A red eastern sky in the morning implies rain drops are starting to grow on condensation nuclei with high humidity to the east. The drops are getting sufficiently large to scatter the longer wavelength rays in the sun's rays. If high humidity and growing drops are to the east, and because weather moves from west to east, the rainy weather would be past us and there would be no need to worry about bad weather and take warning. Similarly, if the red sky is in the west at night, the stormy weather would be to our west and would be coming our way. There should then be no delight if we were to go to sea with the sailor. With this reasoning I decided that the proverb must be wrong.

I then found the proverb in the *Bible,* and I thought that it would be very difficult to imply that the *Bible* may be wrong, so I did some more thinking. In Matthew 16, verses 2 and 3, we find that the Pharisees asked Jesus to show some sign from Heaven to establish His credibility.

The answer given by Jesus in the King James translation is, "When it is evening ye say, It will be fair weather: for the sky is red. And in the morning, It will be foul weather today: for the sun is red and lowering." This is the proverb under consideration stated in Biblical language.

In the *New English Bible* translation, the same passage is given as, "It is a wicked generation that asks for a sign; and the only sign that will be given is the sign of Jonah." There is no statement of the proverb. The New English translation is, according to theologians, the more accurate translation. How could they be so different?

*Bible* scholars find the older manuscripts of the *Bible* do not contain the proverb and that these older manuscripts are worded according to the New English translation. They find further that some scribe did not like the answer given by Jesus in the older manuscript, so the scribe simply added the weather proverb as words of wisdom.

Can the proverb be correct? If the red is in the *west* in the morning and in the *east* in the evening, it would be correct. A further investigation shows that this *can* be the case. Let's consider what can happen in the morning when precipitation is forming to the west and coming our way.

When the sun is just rising, there is an earth shadow that arches across the western sky as a rather dark region. Above the earth shadow in fair weather there is a purple band. The purple is due to blue light and a small amount of red light being scattered back to us with almost all backward scattering. When the western sky contains growing droplets, especially about one micron in diameter, these drops preferentially scatter red light to make the western sky red. Rain would then be coming our way and the sailor should take warning.

There is a similar case of preferential red-light scattering that gives rise to the saying "Once in a blue moon." In this case the region between us and the moon becomes filled with hazy air containing droplets that are about one micron in diameter. And they preferentially scatter the red light from the normally white moon beams. This leaves mostly blue light coming through to make the moon appear blue. Such a natural selection of drop sizes does not happen very often, hence the saying.

I finally decided that the proverb under discussion implied that red should be on the opposite horizon from the sun for it to be true. Another possible explanation that would make the proverb correct is that it was written for a region where there are prevailing easterlies rather than prevailing westerlies. Prevailing winds from the east would make the weather systems move from east to west thus making the proverb true.

In the case of long-range proverbs, where the science of meteorology is much more uncertain, we find more uncertainty in the proverbs. In fact, there are sometimes contradictions such as:

***Rain on Easter gives slim fodder.***

which is just the opposite of:

***A rainy Easter means a good harvest.***

In general, I find most short-range proverbs to have truth hidden in them; long-range proverbs are often quite doubtful. I keep watching long-range proverbs for clues that may give us a better understanding of weather that may occur in the more distant future.

# 9 How Weather Affects People

*The minds of men do in the weather share,*
*Dark or serene as it's foul or fair.*
Cicero

The Roman senate had augurs to observe and interpret omens for guidance in public affairs. One of their duties was to watch the weather before the senate took a vote. Their rule was to look south. If they saw lightning and a storm to their left, the senate should vote. If the lightning and storm was to the right, they should postpone the vote. Because weather moves from west to east, a storm to their right would mean foul weather coming; a storm to their left meant that the storm had passed and fair weather was coming. People feel better in fair weather than in foul weather.

In our progressively more sheltered lives, we tend to be less concerned about how the weather affects us as individuals. We realize how weather affects society as a whole when

*Men work better, eat more, and sleep sounder when the barometer is high.*

there are droughts putting many farmers out of business, or floods driving others from their homes. Usually these events are happening to someone else. They may have some effect on our economic system or they may necessitate a major relief effort. But it is not often that we recognize direct effects on each individual. On the other hand, many of the proverbs do tell us something about direct effects on each of us.

*Never sell your hen on a rainy day.*

*Do business best when the wind's in the west.*

*A coming storm your shooting corns presage,*
*and aches will throb, your hollow tooth will rage.*

*When rheumatic people complain more than*
*ordinary about pain, rain is coming.*

*If corns, wounds, or sores itch or ache more than usual,*
*it will probably rain.*

*If you lick the platter clean, fair weather.*

*When the barometer's low,*
*Teachers say that children misbehave more.*

*If your corns all ache and itch,*
*The weather fair will make a switch.*

With air-conditioned and efficiently heated homes and offices, we are insulated more and more from effects of temperature. This is not true for people who must work outside.

But even in these cases, many outside workers have devised work and play schedules that keep them idle or protected when the temperature is inclement. We cover work areas with plastic sheets and build domed stadiums so we will be more comfortable in winter. Although we shield ourselves from temperature extremes, we cannot shield ourselves from pressure changes.

When air conditioning was installed in schools, it was found that students learned better. Many schools don't have air conditioning because of the short school duration of 180 days over the colder months of the year. But if school terms are increased to 240 days, consideration will have to be given to air-conditioning schools to make the added time yield efficient learning.

At high temperatures, both relative humidity and high temperatures contribute to our discomfort. When the temperature reaches 100F, most people are hot with very low humidity. A temperature of 82F becomes equally oppressive when relative humidity is greater than 70%. Similarly, a temperature of 90F becomes oppressive when the relative humidity is greater than 60%.

The reason for this is our bodies must lose heat as they do work. We can lose some heat by conduction to the air and by radiation, but as the temperature goes up, these exchanges become less and less efficient. The body then depends more and more on evaporation of perspiration. But, as humidity rises, heat loss by evaporation becomes more and more restricted. Body temperature then rises and may lead to heat exhaustion or heat stroke.

The military made studies of radar operators sent from temperate regions to the tropics where both temperature and humidity are high. They found that such operators make many more mistakes as temperature and humidity rises. The operators become more lethargic as time progresses and must be replaced by fresh crews after rather short periods of duty.

Law-enforcement people find that once the temperature

rises above 85F, riots are much more likely to occur, especially when the humidity also starts to rise. Once the temperature exceeds 90F, riot activity subsides, indicating that a certain amount of lethargy develops.

Climatologists have drawn a 70F *isotherm,* a line where the average temperature is 70F around the world. It falls close to the lower latitudes of the temperate zone. They found that most of our civilizations started and thrived close to this temperature region. Civilization spread from this region after people learned how to clothe themselves and grow food, then preserve it in places such as cold caves.

As civilization moved away from the 70F isotherm, people had to learn how to clothe themselves more effectively. In colder climates, people must prevent their bodies from losing too much heat. During World War II, a man by the name of Siple made a study for the U.S. Army in which he measured body-heat loss at various temperatures and under various wind conditions. In order to function, you must have some body-surface area exposed to the environment. Heat loss through this exposed area can be compensated for by eating more food.

Body-heat loss is quite sensitive to the relative wind velocity between the body and environment. Siple found that heat loss in kilocalories per square meter of body surface per hour was given accurately by the formula:

$$H = (10.45 + 10V^{1/2} - 1.0V)(33 - t)$$

In this formula, V is relative wind velocity in meters per second and t is the temperature in degrees Celsius. The 33 gets into the equation because body-skin temperature is about 33C (92F). Everything is compared to marching at 4 miles per hour, or 1.8 meters per second. From this comes a wind-chill temperature.

To use the equation, find wind velocity and environmental temperature and substitute them into the equation to find heat loss. You can then estimate the exposed area of the body for various types of dress and estimate the time you will be exposed to the environment for a given total heat loss. To compensate for additional heat loss, you should eat more than your regular diet.

In the U.S., weathercasters broadcast only wind-chill temperature. In Canada, they also broadcast the heat-loss value, H, which is what you really should know if you're interested in adjusting your dress to the proper heat loss level and compensating for it with a proper diet. The wind-chill index correlates well with the incidence of common colds.

Our bodies contain about 40 liters of fluid. Some of this is blood, much within body cells and some between cells. If we perspire, there must be a transfer of fluids from one of these regions to another. When exposed to cold, more blood comes to the skin surface to keep it warm. There is continuous shifting of body fluids from one region to another, so expect some unconscious reaction to all of these changes.

We make no effort to adjust our immediate environment to atmospheric-pressure changes. As pressure rises, the body absorbs atmospheric gas into body fluids to keep us from being compressed; as pressure falls, we must give off absorbed gases to keep us from tending to explode. Physical laws are such that body fluid can absorb gas a molecule at a time, but when gas is released it first nucleates into small bubbles. These small bubbles can swell body tissue. The release of the bubbles is much like what happens when you open a soft drink that's carbonated.

An extreme case of this phenomenon is the bends experienced by deep-sea divers when they surface too fast. Skin divers become exhilarated as they descend to greater pressure and sometimes descend to their death. Our body fluids—blood, fluid within cells, and fluid around cells—must be continually changing in relative amounts as we sweat to adjust to temperature. These changes are all felt more readily in injured tissue.

With this *asymmetrical* process, expect to feel fine when exposed to high pressure and not so good when the barometer falls. People have different sensitivities to this. There are many examples of how it affects different people. Those with injuries, scars or ailments are more sensitive to pressure changes.

Many a soldier or farmer has lost a limb in battle or in farm machinery. In all cases they feel the pain more when the pressure starts to fall. People with rheumatism complain more when the pressure falls. And school teachers understand how children misbehave more when the barometer falls.

Some teachers with scars or ailments use their increased pain or irritation to tell them to assign more work to keep students out of mischief. Studies have actually been made in pressurized hospital wards to verify that patients suffer more when controlled pressure falls.

Human gestation periods vary in duration about ten days. This variation is a time period about equal to the passing of a weather system. It turns out that most births occur during low-pressure periods: Labor starts when the barometer starts to fall and birth occurs by the time bad weather arrives. Studies show this to be the case. One can perhaps argue similarly that most of us were conceived during high pressure when headaches are at a minimum.

# 10 How Weather Affects Plants

*When corn fodder is crisp, fair weather;*
*When cord fodder is limp, rain is coming.*

*The daisy shuts its eye before rain.*

Plants must interact with the weather in many ways in order to live and grow. Most important is that many become dormant during winter months. While growing in summer, a plant must orient itself properly in the sunlight, it must be sufficiently strong to hold itself up under the force of gravity and respond to the variable forces in all kinds of winds. It must collect water and minerals from the soil and transport them to all parts of the plant. It must be able to adjust to changes in temperature and relative humidity, and it must be able to distribute its seeds.

The plant is continually adjusting to the changing weather. Many of the adjustments are made by rapid differential rates of growth in different regions or by changing the *turgor* (tension produced by fluid) in the cells with various amounts of water.

## Adjustment to Sunlight

The amount of sunlight reaching plants varies greatly, from moonlight at night to intense light from the noonday sun. These variations change with the seasons. Plants are *phototropic* in that they have a response to sunlight that tends to make them grow toward the sun.

Plants adjust to sunlight by orienting their leaves at different angles. There is further adjustment as they orient the chloroplasts within their leaves, changing their green color. The orientation of leaves and flowers and their color will change with the weather.

## Water Transport in Plants

Plants must transport substantial amounts of water from the soil to the air each day. Plants themselves are mostly water, some as much as 90% water, while seeds are as low as 12% water. To transport food and adjust to temperature, a sunflower will transport to the air about one liter of water a day. A birch tree will transport about 60 liters each day.

In order to carry out this pumping process in various percentages of relative humidity, a plant must change the size of its *stomata,* or holes on the bottom sides of its leaves. This is done by changing the *turgor* in the guard cells around the stomata on the underside of the leaf. Plants obtain their rigidity from water under pressure in their cells. When they lose this water by *transpiration,* they are less rigid and we say they *wilt.* The plant looks wilted *or* firm depending on temperature and relative humidity.

An example of how turgor can change the shape of plant structure can be easily demonstrated by picking a dandelion blossom and splitting the ends into four equal parts for a small distance at the bottom of the stem. When it is dipped into water, the inside cells rapidly pick up water and become larger. The four branches then curl outward in a surprisingly short time.

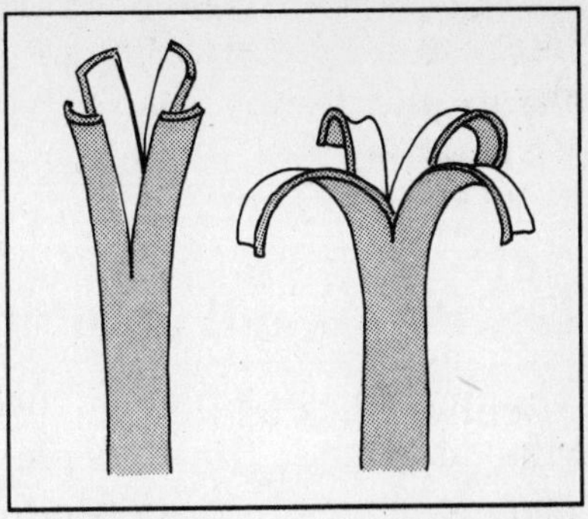

Turgor can be demonstrated by splitting the stem of a newly picked dandelion into four equal parts and dipping it into water. As inside cells pick up water, four branches curl outward.

## Response to Temperature and Other Stimuli

Plants struggle with heat as well as humidity. Much of the cooling in plant life is realized by transpiration where stomata on the undersides of leaves open or close to release more or less moisture to the environment.

***The scarlet pimpernel closes before rain and opens wide in fair weather.***

It is known that the scarlet pimpernel is sensitive to relative humidity. At about 80% relative humidity, its flowers start to close. Some plants respond to temperature changes very rapidly. A good example is the tulip. On a cold day, the blossom is closed. When brought inside with a 10F temperature rise, the flower opens widely in a few minutes. If the flower is placed in a cool refrigerator, it closes again. Crocus flowers are even more sensitive. They will respond to 0.5F temperature changes. The property of a plant responding to temperature change is called a *thermonasty.*

Plants can respond to other stimuli. Such responses are called *nasties.* Some other nasties that can be easily observed are *haptonasties* (sensitivity to touch) in the stamens of flowers and the bushiness of the pistil. When a bee lands on a flower, the flower orients itself so that pollen is transferred. Mosses have several nasties, but they are a little harder to observe because of the small plant size. Sensitive Plants, Pitcher Plants and Venus Fly Traps are further examples of plants that are sensitive to touch.

## Dry Plants and Imbibition

The dry parts of plants, such as seed pods, and products which are made from dry plants, like rope and lumber, show many changes in response to changes in relative humidity. Plants often make use of these changes to disperse their seeds. If a dry pea seed is placed into water, it will greatly increase its volume within a day or so. The reverse of the drying process takes place.

A dry plant is constructed of rather long, thin cells and molecules interlocked loosely together. Water molecules from the air move freely between the air and regions between the cells. As water is picked up from the air, the long plant cells can be completely surrounded by layers of water to make the system acquire the properties of a gel. If the absorption or release of water molecules is nonuniform, plant material can change its shape. A board warping is a good example of this. In seed pods, strong elastic forces can develop so the plant can actually throw its seeds. Learning to recognize various shapes of dry plants can tell you things about relative humidity.

Some plants have developed small outward-facing tubes in their leaves. These tubes are very small and full of water. They have a liquid surface with a very small radius of curvature that allows them to hold liquid water in the tubes even though the relative humidity is much less than saturation. The smaller the radius of curvature, the lower the environmental vapor pressure, so the plant retains its water reserve by changing shape to make the tubes smaller or larger as humidity changes.

Some plants have developed a clever way for opening and closing cavities. Plant cells form a row of cavities similar to a chain of open buckets. Each bucket then becomes partially filled with water. The cavities are small so that the free surface of the water can exercise the effects of surface tension.

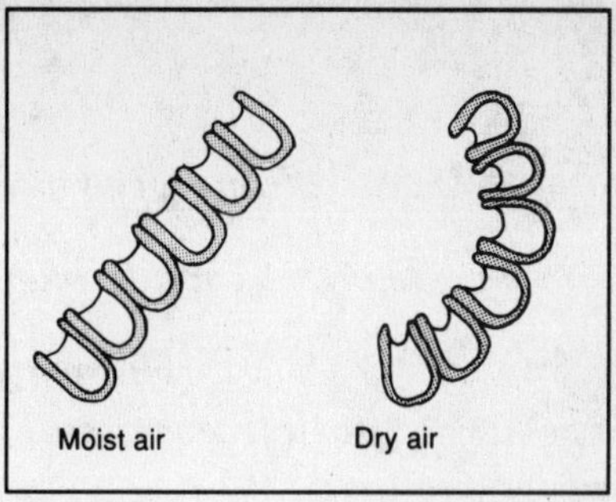


As water evaporates from bucket-shaped plant cavities, more curved meniscus increases surface tension, closing and curling shape of plant structure.

Each liquid surface forms a concave *meniscus*, as shown here. A *convex* meniscus forces the water drop to evaporate more readily and enhance vapor pressure in the air. A *concave* surface will allow liquid to remain in equilibrium with vapor when the relative humidity is much less than 100%. The drier the air gets, the more curved the meniscus becomes. This is realized by the bucket-shaped containers closing and curling the structure. This curling can then change the shape of the structure.

Only a few of the many possible changes that a plant can make have been discussed. If you study the microscopic world of fungi and spores, many more such phenomena are exhibited. Refer to botany books if you have more interest.

## History of Plants and Animals

You should keep in mind that plants and animals have evolved from a liquid environment to an aerial environment on land. The amount of water they gained or lost in a liquid environment was no problem. When plants moved to land, they had to get water from the soil and not lose too much to the air. Plants also had to tolerate much larger temperature changes in the air than those found in water. They also had to adapt structural changes that could hold them in place under the force of gravity and be buffeted by the wind without the benefit of buoyancy forces they had in the water. The changes that they make, if observed, can tell us about the causes of these changes in the weather environment.

✦ ✦ ✦

*Pondweed sinks before a rain.*

*The great white oxeye closes before a rain.*

*The gentian closes*
*both flowers and leaves before a rain.*

*The sudden growth of mushrooms presageth rain.*

*Purple sandwort expands its beautiful pink flowers*
*only when the sun shines,*
*but closes them before a coming shower.*

*Clover contracts its leaves before a storm.*

*When the down of a dandelion contracts,*
*it is a sign of rain.*

*Goat's beard keeps its flower closed before rain.*

*Sensitive plants contract their leaves*
*at the coming of rain.*

*Cottonwood turn up their leaves before a rain.*

*Pine cones will close in wet weather*
*and open in dry weather.*

*Dead branches falling in calm weather*
*indicate rain.*

# 11 More Proverbs

***Corn is as comfortable under snow as an old man is under his fur coat.***

Snow on the ground insulates the ground from deep frost. Plant roots are not harmed by freezing and water can be retained by the soil when the snow melts.

***Much snow much hay.***

***The more snow, the more healthy the season.***

***If the first snow sticks to the trees, it foretells a bountiful harvest.***

***When snow falls dry, it means to lie;***
***But flakes light and soft bring rain oft.***

Dry snow forms high in clouds during very cold weather. Soft fluffy snow is formed near the cloud base at much warmer cloud temperatures. A slight increase in temperatures will then make it turn to rain.

❄

***As many days old as the moon is at the first snow, there will be as many snows before crop planting time.***

***The number of days the last snow remains on the ground indicates the number of snowstorms which will occur during the following winter***

These long-range proverbs seem to make little sense. I can think of no meteorological reason why they should be true.

❄

***In frosty weather, the stars appear clearest and most sparkling.***

Frosty weather and bright stars indicate stable air with little moisture to absorb outgoing radiation.

❄

***Hail brings frost in the tail.***

***A hailstorm by day denotes a frost at night.***

In the early fall, a cold front may pass. Hail is produced in severe cold-front storms that result from quite contrasting temperatures across the front. The cold front is followed by cold, dry air in a building high-pressure system that can lead to frost.

***Rain before seven, clear by eleven.***

***The sudden storm lasts not three hours.***

***The sharper the blast***
***The sooner 'tis past.***

These proverbs describe the passing of a cold front in which a storm builds quite rapidly and passes in two to four hours. The frontal system is about 200 kilometers (125 miles) wide where tall cumulus clouds develop and travels at an average speed of 50 kilometers per hour (31 miles per hour).

***Mony (many) hips and haws,***
***Mony frosts and snaws.***
Scottish

Hips are the ripened fruit of wild roses, and haws are the berries of the hawthorn tree. This long-range proverb says that a good growing season for these plants indicates a severe winter. The more berries could reflect good growing weather in the summer or possibly lots of pollination by the activity of bees or other insects.

***The sun is hotter when it shineth forth between clouds than when the sky is open and serene.***
Bacon

This is a deception to our senses. When you dip a finger into tepid water after holding it in cold water, it feels warm. But after being held in warm water, the finger feels cooler.

***The sun getting up its back-stays indicates foul weather.***

Many people refer to this phenomenon as the sun drawing water. The clouds have a sufficient number of openings to define pencils of light. Within these pencils of light, water drops have grown sufficiently large to scatter light, preferably blue light to make the stays have a bluish hue.

***The heat of beams of the sun doth take away the smell of flowers, especially those of milder odor.***

Bacon

In hot, dry weather, aromatic molecules from flowers become less hydrated and do not bind as well to the nasal skin where chemical action can lead to the sense of smell.

***No weather is ill***
***If the wind is still.***

Much of the wind is caused by the release of heat energy in the clouds. Clouds serve as heat engines that drive the atmosphere. If there are no clouds to make the wind, the weather is fair.

***Unsteadiness in the wind shows changing weather.***

***A frequent change of wind, with agitation in the clouds, denotes a storm.***

The unsteadiness shows that the air is becoming more unstable so that clouds can start to form.

***A whispering grove tells of a storm to come.***

***Whistling telephone lines tell of a storm to come.***

These are aeolian winds. An object in the wind's path puts trailing vortexes (whirling masses of air) into the wind stream. Telephone lines whistle louder in cold weather because the lines are tighter. Oak trees create a much greater range of size in these trailing vortexes to give more varied pitches to the sound. W. J. Humphreys in his book *Physics of the Air* describes how the muffled complaint of the oak is quite different from the hissing sigh of the pine.

***The north wind makes men more cheerful,***
***and begets a better appetite to meat.***

***Fair weather cometh out of the north.***

***A northern air brings weather fair.***

The wind usually shifts to the northwest or to the north when a high-pressure system moves in with clockwise circulation about its center. We feel better during high-pressure periods.

***An honest man and a northwest wind***
***generally go to sleep together.***

***The west wind is a gentleman and goes to bed.***

Most winds tend to decrease at night at the ground, although clouds can still be moving above. At night ground heating vanishes so there is no turbulent mixing between the ground and regions above. Frictional coupling between upper air and the ground is then less, and ground air is not dragged along by upper air wind at night.

***When distant hills are more than usually distinct, rain approaches.***

***When cliffs and promontories on distant shores appear higher, then the southeast wind is blowing.***

There are two effects being described. The air is becoming unstable and light-scattering centers are raised to form a cloud. The air is becoming unstable due to heating at the ground. This leads to an inferior mirage that makes things appear taller. Hot air at the ground becomes buoyant, or unstable, and rises. A southeast wind is usually bringing in moisture from a warm body of water to the south.

***A good hearing day is a sign of wet.***

***Sound traveling far and wide,***
***A stormy day will betide.***

Both heating of the air near the ground and vertical wind shear can refract sound waves to give sound mirages. Sounds going against the wind are refracted upward while sounds with the wind are refracted downward. Sounds are heard better to windward when heard in elevated listening posts such as the crows' nest on a sailing ship.

Bells are placed in high steeples so they can be heard

farther away. Sounds that go to windward may not be heard directly, but their echoes from a cliff may be heard clearly. The wind and heating of air near the ground are signs of developing instabilities.

***When a rainbow appears in the wind's eye,***
***rain is sure to follow.***

***Rainbow to windward, foul falls the day;***
***Rainbow to leeward, damp runs away.***

The rainbow is due to refraction and reflection of sunlight by raindrops. With these raindrops to windward, the storm will surely move our way.

***A rainbow in the east will be followed by a fine morrow,***
***in the west by a wet day.***

***A dog in the morning, sailor take warning;***
***A dog in the night is the sailor's delight.***

A *dog* is sailor talk for a small rainbow near the horizon. Because weather systems move from west to east, any indication of raindrops to your west will indicate that rain is coming, and anything that indicates rain to your east implies the rain is past.

***None so surely pays his debt***
***As wet to cold and cold to wet.***

This reflects the passing of low-pressure systems with storms followed by high-pressure systems with fair weather.

*Goats graze down the mountain before a rain and up the mountain for fair weather.*

*Goats leave high ground and seek shelter before a storm.*

*When sheep turn their backs to the wind, it is a sign of rain.*

*Sheep fighting for their food more than usual indicates a storm.*

*The goat will utter her peculiar cry before rain.*

It appears that sheep and goats are sensitive to a falling barometer when gases are escaping from their body fluids.

*If oxen turn up their nose and sniff the air, a sign of rain.*

*If oxen lick their hair the wrong way, a sign of rain.*

*If oxen lick their front hooves, it is a sign of rain.*

*If horses stretch their necks and sniff the air, a sign of rain.*

*If dogs roll on the ground and scratch or eat grass and refuse meat, a sign of rain.*

*The unusual howling of dogs portends a storm.*

All of these seem to indicate a form of irritation that may be caused by a falling barometer when absorbed gases escape from body fluids and collect as small bubbles in body tissues.

*When the cat lies on its brain,*
*Then it is going to rain.*

*When cats place their paws over their ears,*
*it is a sign of rain.*

*When cats hide under the bed,*
*there will be a storm.*

Cats can become irritated by falling pressure and do things that indicate itchy skin. However, they also lick themselves, usually in the direction of their hair, in fair weather. This removes electrostatic charges accumulated during fair weather and low humidity. Cats have a keen sense of smell, and may respond to smell changes when large molecules become hydrated with water molecules as humidity rises.

*If cows refuse to go to pasture,*
*expect a storm.*

*If dogs howl, expect a storm.*

*If bulls are irritable, expect a storm.*

*If a cow thumps her ribs with her tail, expect a storm.*

*If rats and mice are more restless,*
*expect rain.*

*If moles throw up more earth than usual,*
*expect rain.*

Most of these indicate a form of irritation due to a falling barometer. Remember that people are also more irritable during low-pressure periods.

*Porpoises in a harbor, expect a storm.*

*When dolphins and porpoises play near a ship, it is a sign of a storm.*

These animals must come to the surface to breathe. Apparently this is more difficult in rough water. A ship's wake often leaves patches of calm water. Sailors say dolphins around their ship is a bad omen.

*When fish bite readily and swim near the surface, expect rain. They stop biting just before a thunderstorm.*

*Cuttle fish swim near the surface before a rain.*

*Fish swim upstream and catfish jump from the water before a rain.*

A falling barometer releases nymphs and other food from the bottom by means of expanding more buoyant bubbles. This starts the food chain in progress. Why they stop biting just before a thunderstorm is difficult to determine because electric fields before a lightning strike can't penetrate water. But currents from lightning can penetrate water and kill fish just as they kill animals on land. The spreading of lightning current in fresh water is much greater than on more conducting land. You should get off the water when a thunderstorm approaches simply because a higher object on a lake surface is much more likely to be struck.

*If snails are abundant, it is a sign of rain.*

*When black snails cross your path,*
*Black clouds much moisture hath.*

*When black snails on the road you see,*
*Then on the morrow rain will be.*

*When snails crawl up an evergreen*
*and remain there all day, expect rain.*

It appears that snails require humid air in order to be away from water or moist environments. The high humidity can likely lead to rain.

*When frogs warble, they herald rain.*

*The louder the frog, the more the rain.*

*When frogs spawn in the middle of water, it is a sign of drought; and when at the side, it foretells a wet summer.*

Frogs like snails require high humidity in order to be out and about. How they can detect where the water level will be when the spawn hatch is unknown.

*If spider frames are long, they predict fair weather.*

*If spiders are indolent, rain will soon come.*

*When spiders build new webs,*
*the weather will be clear.*

*If spiders leave their webs, expect rain.*

Spider webs are very sensitive to moisture and contract when wet. This contraction destroys the web. You will see long filaments only when the weather is dry.

*The early appearance of insects indicates an early spring and good crops.*

*If ants their wall do frequent build,*
*Rain will from the clouds be spilled.*

*When ants migrate away from low ground, it forbodes rain.*

Ants may be able to sense an increase in relative humidity by more liquid water condensing in their very small tracheal tubes. In migration they travel in lines following the trail of a leader who releases a pheromone for other ants to follow by smell. The ability to smell is enhanced by the pheromone molecules being more hydrated when humidity is high.

*When eager bites the thirsty flea,*
*Clouds and rain you sure shall see.*

*A fly on your nose, you slap, and it goes;*
*If it comes back again, it will bring a good rain.*

*If flies cling to the ceiling, rain may be expected.*

*Flies tend to collect in swirling swarms before rain.*

I think flies sense relative humidity through collection of liquid water in their small tracheal tubes. I'm not sure why they are more likely to bite. Tests were made with powerful radar sets to see if air turbulence could be detected before a rain cloud formed. Testers got very strong radar reflections and found them to be swarms of swirling flies.

*Fall bugs begin to chirp six weeks before a frost in the fall.*

*When locusts are heard, dry weather will follow, and frost will occur in six weeks.*

All I know about this is that many people believe it.

*One can tell the summer's weather by observing the cocoon of the cicada; if the head is upward, a dry summer; if the head is downward, a wet summer.*

If true, this is quite mysterious.

*When crickets chirp exceptionally, wet is expected.*

*To tell the temperature, take 72 as the number of chirps per minute at 60F; for every four chirps extra, add 1 degree F; for every four chirps less, subtract 1 degree F.*

Crickets are cold-blooded and their activity is proportional to the temperature.

*If the clock beetle flies circularly and buzzes, it is a sign of fair weather.*

These beetles were sacred to the early Egyptians who thought them to be symbols of truths that are too deep for words. They carved many scarab-beetles in their monuments as symbols of immortality.

***If the sun sets behind a cloud,***
***it forebodes rain the next day.***

***When the sun sets in a bank,***
***A westerly wind we shall not lack.***

***Glimpse you ever the Green Ray,***
***Count the morrow a fine day.***

These proverbs tell us that there is moisture to the west and it is coming our way. It may be a cold front that will shift the wind to the northwest when it passes. The Green Flash is sometimes seen when the sun is at the horizon over a large body of water. Green rays are preferentially scattered or refracted to you. It is said that every girl should see this before marriage if wedded life is to be a success.

***When the moon's outline is not clear,***
***rain is to be expected.***

***If the moon shows a silver shield,***
***Be not afraid to reap your field;***
***But if she rises haloed round,***
***Soon we'll tread on deluged ground.***

***When the wheel is far, the storm is n'ar;***
***When the wheel is near, the storm is far.***

***Double circles around the moon***
***portray very severe weather.***

These proverbs relate to refraction of light by ice crystals very high in the sky, which is moisture blown from the tops of tall thunderstorms coming our way. The first halo at a 22-degree angle is due to refraction by six-sided columnar crystals of ice. A second halo occurs when the crystals are more pointed on the ends, leading to more complicated light paths through the crystals.

*Aurorae are almost invariably followed by stormy weather in from 10 to 14 days.*

*The aurorae, when very bright, indicate approaching storms.*

Modern studies show that after aurorae, there tends to be an increase of cirrus clouds over the North Pacific Ocean and an enhanced vorticity or turbulence in the weather system generated at that time.

*Last night I saw St. Elmo's stars,*
*With their glimmering lanterns, all at play*
*On tops of the masts and tips of the stars,*
*And I knew we should have foul weather today.*

St. Elmo's fire is a corona discharge due to the strong electric field of an approaching thunderstorm. When the phenomenon is seen, the storm usually strikes within the hour and not on the 'morrow.' The discharge is ghostly in appearance. One discharge point was called *Castor;* two were called *Castor and Pollux.* One meant a bad storm, two meant that there would be good luck surviving the storm and more than two meant a very severe storm. Some believe that the burning bush described in the *Bible* was St. Elmo's fire.

*The winds of the daytime wrestle and fight,*
*Longer and stronger than those of the night.*

*The lighter winds rise in the morning and fall at sunset.*
Bacon

*Greater wyndes chaunce in the daye than in the nyght.*

Late Middle English, 1555

*Winds which begin to blow in the day last longer and have more force than those which begin to blow at night.*

Theophrastrus

Coupling between upper winds and those at the surface depends on vertical mixing of air heated at the ground by the sun. This heated air becomes unstable and, in doing so, increases friction drag. After sunset this heating stops and there is less mixing, so the wind dies away even though the clouds above continue to move.

*When the wind backs and the weather glass falls,*
*Then be on your guard against rains and squalls.*

*If the wind follows the sun's course, fair weather.*

As viewed from above, wind direction can change with time either in a clockwise (veering) direction or in a counter-clockwise (backing) direction. The sun passing across the southern sky from east to west is veering. When wind veers with time, a low-pressure center is passing from west to east to your north. When wind backs with time, the low-pressure center is passing from west to east to your south. In the backing case, you are likely to be caught in the head of the large comma-shaped cloud-cover regions seen in satellite pictures of weather systems.

*In by day and out by night.*

This describes land breezes and sea breezes. During the day, the land absorbs more of the sun's heat so that air over land is buoyed upward to be replaced by air from the sea. At night, the land cools faster than the water so that air over the water is buoyed upward and replaced by land air.

*If small birds seem to duck and wash in the sand,*
*it is a sign of coming rain.*

*If the fowls huddle together outside the henhouse instead*
*of going to roost, there will be wet weather.*

*If fowls grub in the dust and clap their wings,*
*or if their wings droop, or if they crowd into a house,*
*it indicates rain.*

*If fowls roll in the sand,*
*Foul weather is at hand.*

I think that the birds are experiencing irritation due to a falling barometer.

*If cocks crow during a downpour,*
*it will be fine before night.*

*Land birds bathe before a rain.*

*Geese call down the rain with their cackling.*

*If ducks and geese fly back and forth and duck into the*
*water to wash themselves, wet weather will ensue.*

*Parrots whistling indicate rain.*

*If the wild geese gang out to sea,*
*Good weather there will surely be.*

*The goose and the gander*
*Begin to meander;*
*The matter is plain,*
*They are dancing for rain.*

*Turkeys perched in trees and refusing to descend*
*indicate snow.*

*When swans fly, it is a sign of wet weather.*

*If peacocks cry in the night,*
*there is rain to follow.*

*If pigeons return home slowly,*
*expect rain.*

*Birds in the lowland predict snow;*
*Birds in the highland expect fair weather.*

*When crows go to the water and beat it with their wings,*
*expect foul weather.*

*If woodpeckers are much heard,*
*expect rain.*

*A screeching owl indicates a storm.*

*If sparrows chirp a great deal,*
*expect a storm.*

*Petrels gathering under the stern of a ship*
*indicate foul weather.*

*If starlings or crows congregate,*
*expect rain.*

Birds most likely sense the falling barometer through irritation or by changing air density. I think they can also hear low frequencies with their feathers as low-frequency sound detectors. They can detect air turbulence we can't hear. Petrels are birds that always seem to be flying far at sea. Before storms, sailors noted that they would congregate around a sailing ship for protection.

❄

*Short boughs, short vintage.*

*Plenty of berries indicates a severe winter.*

*When oak trees hang full, expect a severe winter*
*with much snow.*

*If the oak bears many acorns,*
*it foreshows a long and hard winter.*

*When the bramble blossoms early in June,*
*an early harvest is expected.*

*Frost will not occur after the dogwood blossoms.*

*If maize is hard to husk,*
*expect a severe winter.*

*Abundant wheat crops never follow a mild winter.*

*If many dog-roses are seen,*
*expect a severe winter.*

*Dead nettles in abundance late in the year*
*are a sign of a mild winter.*

*When the mulberry buds and puts forth leaves,*
*fear no more frosts.*

*When the onion skin is thin and delicate,*
*expect a mild winter;*
*but when the bulb is covered by a thick coat,*
*it is held to foreshow a severe season.*

*If the feathers of water fowl be thicker and stronger,*
*expect a severe winter.*

These are all long-range proverbs. Although one can find no good reason for them being true, it may be worthwhile to keep them in mind to serve as clues on how other organisms sense the weather.

# 12 Proverbial Finale

Much of our proverbs work I can summarize with a poem. I would like to credit the author, but as with many of the proverbs we've studied, the poem's authorship is lost in time.

## Animal Behavior

*When the small birds prune the wing,*
*Duckling in the limpid spring.*

*Languid 'neath the sheltering trees*
*Oxen snuff the southern breeze.*

*Cackling geese with outstretched throat*
*Join the crow's discordant note.*

*Busy moles throw up the earth,*
*Crickets chirping on the hearth.*

*Loudly caws the harsh toned rook,*
*Spotted frogs respondent croak.*

*Gnats whirl round in airy ring,*
*Angry wasps and hornets sting.*

*Cautious bees forbear to roam,*
*Honey seeking near their home.*

*Spiders from their cobwebs fall,*
*Forth the shiny earthworms crawl;*
*Loud, sonorous asses bray.*

*Frequent crows the bird of day,*
*Hens and chicks run helter-skelter.*

*These though cloudless be the day,*
*Tokens are that rain is nigh.*

# Meteorology

*Who has seen the wind?*
*Neither you nor I,*
*But when the trees bow down,*
*The wind is passing by.*
Christina Rossetti

# 13 Introduction to Meteorology

Throughout my previous discussion of weather proverbs, I showed how each proverb relates to some parameter that helps describe the weather. You may or may not know how weather parameters fit together to describe a weather system. The following chapters describe in more detail how a particular parameter fits into the general picture of the weather.

I show what relative humidity tells us about the condition of the air and how air with high humidity is more likely to lead to rain. I'll discuss how rain drops form and how the condensation process leads to the release of energy which can set the atmosphere into motion.

There is a chapter on how air masses interact to give us warm and cold fronts that are always of concern. Air motion on a rotating earth leads to some unexpected relations between pressure differences and wind directions.

I show that when frictional forces of the wind are near

the ground, vertical motions are necessary. There is downward motion in high-pressure regions and upward motion in low-pressure regions. These processes lead to precipitation and storms in low-pressure regions where the energy of condensation can be released. I also show why fair weather is associated with high pressures and foul weather is associated with low pressures.

Carefully observing the details of air movement and rising at fronts will help you predict the weather for a couple of days. Fronts have many different characteristics. Once you've identified them, they will be a great aid in predicting future weather events.

The types of clouds and their movements above us also help us predict warming or cooling trends in the weather.

There is always some subtle way of partially extracting some of the weather information that professionals get with the use of sophisticated instrumentation. Because everyone other than practicing meteorologists gets most information about weather by simply watching and listening, I include a short section on how sunlight's interactions with clouds and raindrops can tell us something about the meteorology of any weather situation.

# 14 Relative Humidity

## Dance of the Water Molecules

Relative humidity is a quantity, typically expressed as a percentage, that tells us the amount of water vapor in the atmosphere under certain conditions determined by temperature. Actually, relative humidity has nothing to do with the fact that there are many *other* kinds of molecules in the atmosphere. Relative humidity is a measure of the exchange between water molecules in liquid-water form and those in vapor form.

Relative humidity may be easier to understand if you think about it from the standpoint of molecular motions. Atmospheric air consists of air molecules and water molecules moving very fast in a chaotic way. If there is a surface of water in liquid form, the air molecules will strike this surface and bounce away. On the other hand, water molecules *in* the air moving from the air to the liquid surface will stick to the surface.

If this were all that happened, all water molecules in the air would end up in liquid form. But water molecules in liquid form are also in motion, and the speed of this motion depends on the temperature. In some of the molecular collisions between water molecules in the liquid near the surface, some molecules may receive sufficient energy to escape from the surface. The chance for this escape doubles for about every 10C (18F) increase in temperature. We call this process *evaporation.*

Visualize water molecules streaming *toward* the liquid surface from the air where they are captured, while other water molecules are *escaping* from the surface and streaming *back* into the air. If the number of water molecules entering the liquid surface in a given time equals the number escaping, the air is saturated and relative humidity is 100%.

When relative humidity is close to 100%, water returns to liquid form from the air as fast as it leaves, resulting in negligible evaporation. Anything wet remains wet. If there is to be a net condensation, such as forming droplets in a cloud, relative humidity must be greater than 100%. If fewer molecules stick to the liquid surface than leave, the air is *unsaturated* and liquid water is evaporating. The surface is losing more molecules than it gains. The ratio, expressed as a percent, of the number of molecules entering the surface to the number leaving is *relative humidity.* If the percent is small, many more molecules leave the surface than enter, and the liquid water will evaporate. Clothes on the washline dry rapidly.

The number of water molecules leaving a liquid surface depends on temperature. Relative humidity is a ratio of the number of incoming molecules to the number of outgoing molecules. So it should be obvious why the same value of the relative humidity at different temperatures leads to different amounts of water vapor in the air.

A given relative humidity on a hot day means there is much more water vapor in the air than for the same relative humidity on a cold day. When the temperature is high, many more water molecules can escape from the liquid. So the atmosphere then must have much more vapor to get the same flow of vapor back to the liquid.

Air at high temperature and high humidity contains lots of water vapor. This is why we tend to get more precipitation (rain) in the summer than in winter.

## Relative Humidity and Dew Point

Some common examples of how relative humidity affects us may help in understanding definitions. Consider sweat on a glassful of cold iced tea on a humid summer day.

The flow of streaming water vapor to the glass surface is obviously greater than the rate of escape, so liquid water drops form. Return flow from the cold drops on the glass is small because the contents of the glass keep the temperature low. Even though surrounding air may have a relative humidity of less than 100%, it can furnish sufficient moisture flow to the glass to cause condensation. If the surrounding air is very dry, it may not be able to supply water molecules fast enough and there will be no condensation.

If moisture flow to and away from the surface are equal, the surface is at the *dew-point* temperature of the air. A meteorologist determines dew-point temperature by cooling a surface until condensation just starts to form. From this he can calculate the flow of vapor to and from the surface and give us a value for relative humidity.

## Seeing Our Breath

In the winter we can see our breath, but not in summer. Our body temperature is the same all year and exhaled air from our lungs is nearly saturated with water from our bodies at all times. Breathe onto a mirror that is at a temperature lower than body temperature and it will fog. This is because the flow of water vapor to the surface is greater than the flow away at mirror temperature. The colder the mirror, the easier it is to make it fog. If the mirror is above body temperature, it will not fog.

On a cold winter day, we don't need a mirror to see our breath because condensation takes place on very small aerosol particles in the air. Normally these are too small to see, but as they collect moisture they grow to visible size. The drops are still quite small and their surfaces are curved.

Even on cold days, visible drops again evaporate. This happens because of an effect known as *surface tension.* Liquid molecules attract each other and, as they merge, they contract into spheres, the geometric shape which has the smallest surface area. These forces continue to act to make the surface smaller and smaller and squeeze the drop into oblivion. There is then a struggle between vapor trying to condense on the aerosol particle and surface-tension forces squeezing the drop to death.

In summer the condensed drops cannot grow to visible size before they evaporate. A smoker exhaling can see his or her breath in winter or summer because the beginning aerosol particles are much larger.

The presence of small drops being forced to evaporate rapidly by surface-tension forces can raise relative humidity above 100%. The smaller the drop, the more the increase. Consider a large drop and a small drop in the same region. Relative humidity can become of such a value that the small drop evaporates due to enhanced surface-tension forces. Also, the large drop can grow due to the enhanced relative humidity. It is by this process that rain can form. Large drops steal water from small drops until they become large enough to fall as rain.

## Forming Cloud Droplets

The process by which rain forms can be demonstrated with a simple experiment. A large, clean glass jug with a fraction of a cup of water at the bottom is allowed to stand at room temperature until the air in the jug becomes saturated with vapor. If you have a tube through the cork in the jug, you can suck on the tube and make the air in the jug expand a little. This expansion cools the air in the jug, just as air rushing from a tire produces cooling. This cooling now makes air in the jug become supersaturated at the lower temperature. If the jug has been standing a while, the aerosol particles will have drifted to the jug wall so that there is nothing on which the supersaturated vapor can condense. A match is then lighted and held over the tube so that a little smoke can drift into the jug. If you now suck on the tube, a dense fog will appear in the jug. Smoke furnishes aerosol particles on which water condenses, making droplets.

## Energy Transfer—Why Snow Melts Faster on Humid Days

Perhaps you have noticed how snow melts exceptionally fast on certain humid spring days. On these days the flow of molecules to ice and snow surfaces is much greater than the flow away from them. A great deal of water vapor then moves from the air and is captured on the snow surface. Rapid melting occurs because each condensing molecule gives up a large amount of energy as it condenses. Each molecule of condensing vapor has sufficient energy to release about seven water molecules from an ice crystal to form more loosely-bound liquid water. This results in rapid melting of the snow.

Molecular motion is always altered when vapor goes to liquid or liquid goes to solid. And it requires considerable energy to alter this motion. Energy is given up during condensation and must be provided during evaporation.

Meteorologists make use of this energy exchange when measurements are made with *wet-bulb* and *dry-bulb thermometers.* A dry-bulb thermometer is an ordinary thermometer. A wet-bulb thermometer has a water-soaked bag around the bulb. When a wet-bulb thermometer is whirled in the air, water evaporates from the bag and lowers the temperature reading as heat energy is taken from the thermometer to evaporate the water. The amount of cooling is related to the dryness of the air.

## The Greenhouse Effect

The amount of water vapor in the air is not very much, even if the air is saturated. The maximum amount is about 4% by volume. It is this small amount of water that must condense in clouds to produce rain. Clouds form when the air is cooled to its dew-point temperature. Even when there are no clouds, there will be invisible water vapor in the air. These vapor molecules can absorb or release energy by changing their *rotational* and *vibrational energy*.

On cool nights, soil and vegetation radiate daytime energy back to space, which is why the earth doesn't get hotter and hotter from solar energy. Some of this radiated energy can be caught by water molecules in the air, changing their rotation and vibrational motions. These molecules can lose this energy by radiation. But some of the energy is radiated back toward the earth's surface, keeping the surface considerably warmer than it would be if water vapor were not in the atmosphere. This is known as a *greenhouse effect.* When there is cloud cover, this effect is greatly enhanced so cloudy nights are most often much warmer than clear nights. The higher the dew-point temperature, the less chance there is of frost.

# 15 Motions of the Atmosphere

## Physical Properties of the Atmosphere

The motion of the atmosphere on the rotating earth tends to be a little complicated. If we first consider a few simple rules, it will be easier to understand.

Air is a gaseous fluid. This fluid property means that it can take any shape, fit into all crevices or, in general, take on the shape of any container. As a gas, its volume is limited only by the volume of the container. A liquid, on the other hand, is a fluid that will occupy a definite volume or part of its container, but not necessarily fill the container.

The atmosphere is somewhat different than most containers. The earth's surface serves as one boundary, but there is no obvious second boundary at the top. The upper surface of the atmosphere is infinitely diffuse, and is formed by the pull of gravity.

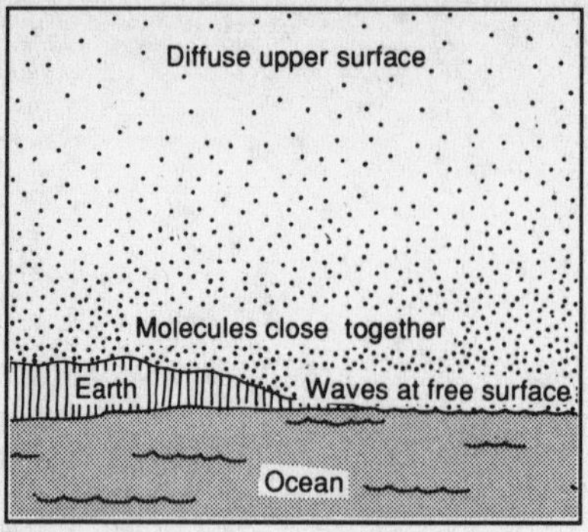


Molecules in upper parts of the atmosphere are spaced farther apart than those at lower levels, which are squeezed close together by the weight of the air above.

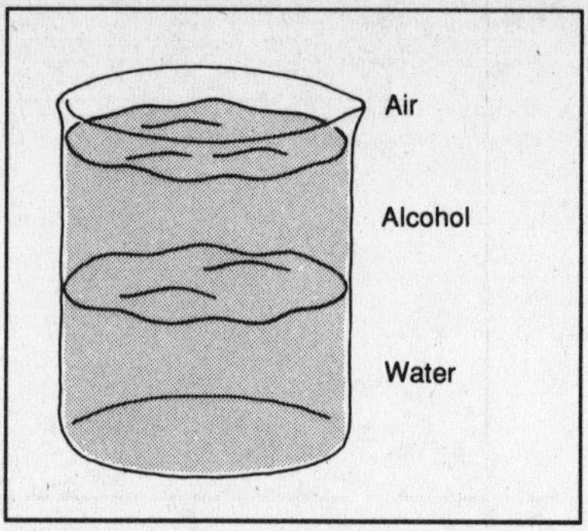


Because water is more dense than alcohol, it goes to the bottom of the container. Air flows to the top. Waves can exist at the interface of any two adjoining liquids.

Gravity pulls back air molecules that try to escape from the earth's atmosphere. But there is no well-defined upper limit to the height (altitude) where this happens. I contend that there is no upper free surface for a gaseous atmosphere. Air molecules in the very uppermost parts of the atmosphere carry out motions that are more like projectiles shot from a gun that rise into space then fall back due to the force of gravity. Their motions eventually blend in with similar but more complicated motions of the sun's atmosphere. Air in the lower parts of our atmosphere is always compressed by the weight due to gravitational force on all of the air above.

The oceans are also fluids, but they are liquid rather than gaseous. As a liquid they occupy a fixed volume and, hence, have a very definite upper free surface. Interactions between the upper surface of the liquid ocean and the atmosphere leads to an ever-changing shape in this free surface that we see as *waves.* In addition to waves in the ocean/atmosphere interaction, there is a large exchange of water vapor as explained in Chapter 13. And there is a large exchange of energy in the form of heat and mechanical energy which can lead to strong currents in both fluids.

If there are two fluids of different density in a container that are not mixed, the denser fluid will settle to the bottom. For example, if water and alcohol are placed in a vessel in such a way as to avoid mixing, the denser water will be at the bottom of the vessel. In this case, there are two fluid surfaces; one between the liquid water and liquid alcohol, and a second between the liquid alcohol and the gaseous air. Waves can be set up in either of these surfaces.

## Initiating Motion in the Atmosphere

You may think that because there is only air in the atmosphere, fluids of different density are of no consequence. This is not the case because different air masses may be at different temperatures and pressures. Both factors change air density according to the following: *Higher pressures tend to increase air density; higher temperatures tend to decrease density.*

When adjoining air masses are subjected to different pressure and temperature changes, the air masses will have different densities. Due to gravity, the denser mass will tend to move under the less-dense mass. In the gravitational field of the earth, the pressure forces are determined mostly by the force of gravity. So temperature variations become an important factor in determining stirring motions of the atmosphere. Air masses with different amounts of heating can rise and fall in the atmosphere.

Motions of the earth's atmosphere become complicated because most of the solar energy enters near the equator. All of this energy is radiated back to space more or less uniformly from all points on the surface of the earth. This means that solar energy must be distributed by motions of the atmosphere.

Because most energy comes to us at the equator, it is not difficult to see why surface temperatures are higher at the equator than at the poles. It is a little more difficult to see why temperature decreases with altitude in all regions. We

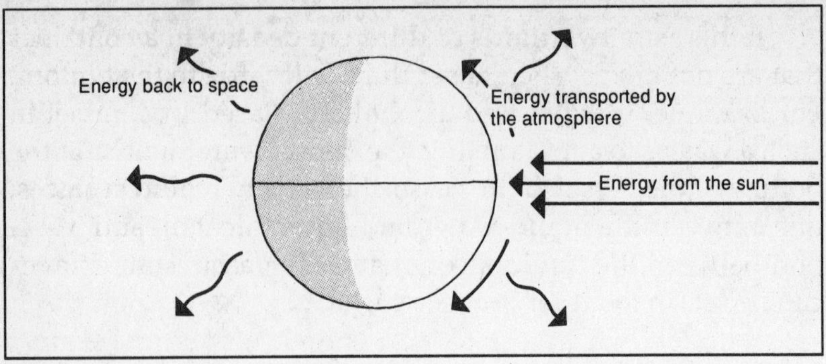


Solar heating occurs mostly at the equator and is lost mostly through radiation. This loss is equally distributed over the earth's surface.

observe that most sunlight can pass through the transparent atmosphere. But there can be what is called *infrared heating* and *ultraviolet heating* by radiation that can't be detected by the eye.

Our eyes have adapted to visible radiation that can readily pass through the atmosphere. Infrared and ultraviolet radiation can deposit considerable amounts of energy directly into the atmosphere. This is realized by ozone absorption of ultraviolet light at high altitude. Similarly, absorption of infrared by carbon dioxide and water vapor occurs in varying amounts at lower altitudes.

I have argued that the atmosphere must circulate in order to distribute energy that enters mostly at the equator. In this circulation, air rises and falls between upper and lower regions. If we take air up in one region, it must come down in another.

As any air moves upward, the weight due to the air above becomes less and less so pressure decreases. This decreasing pressure allows the air to expand to a larger volume, reducing its density. In the process of expanding, it affects its surroundings. The air uses the energy created by its molecular motion, which we measure as "temperature," to do this work.

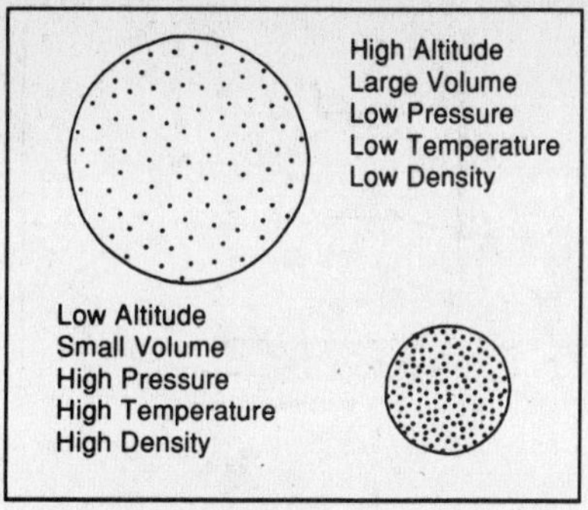

Air pressure, temperature and density vary inversely with altitude; volume varies directly. In other words: the higher the altitude, the higher the air volume, but the *lower* the air pressure, temperature and density.

As rising air does work on the surroundings, it loses its internal energy or some of its molecular motion to make the temperature decrease. Conversely, if the air descends, the temperature must rise as the surrounding air does work on it to squeeze it into a smaller volume. Therefore, there is a continual vertical exchange of air masses with higher temperatures near the surface and lower temperatures at high altitudes.

The amount of temperature change as we go up and down in the atmosphere also depends on the starting temperature, so the situation is somewhat different at the poles compared to that at the equator. These differences are all that are needed to give the atmosphere its general circulation of air rising at the equator and sinking at the poles. At the poles, air can descend and move toward the equator to replace the air that has moved upward. The detailed motion is more complicated than this, but these are the guiding motions that transport excess energy from the equator to the poles.

Because we observe most of our weather from the earth's surface, it will generally be dominated by a motion of cold air from the polar regions toward the equator. Meteorologists say the atmosphere behaves as a giant heat engine, distributing heat and moisture as these motions are generated. In setting up the motions, interactions with the oceans result in ocean currents, which also distribute the energy brought in at the equator.

Masses of colder air leave the polar regions and move toward the equator next to the earth's surface while warm air near the surface in equatorial regions is forced upward and

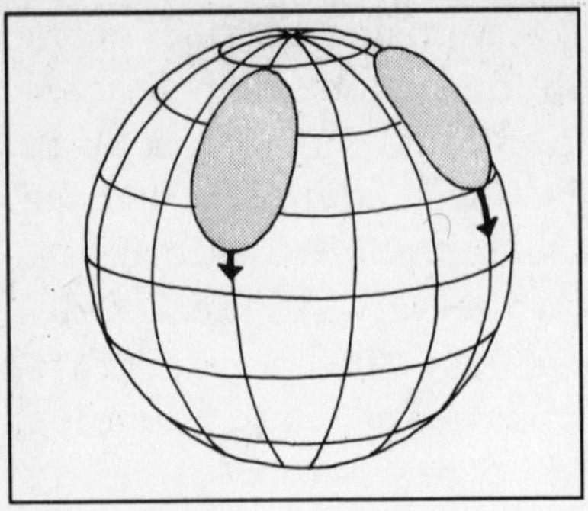

Warm air flows from the equator to the poles and returns to the equator as bubbles (fronts) of cold air wedged under the warm air.

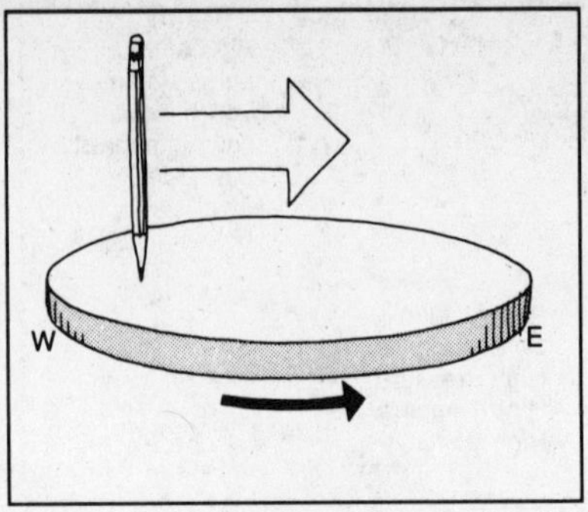


Simulate motions of cold-air bubbles by moving pencil across disc that's rotating west to east, or counterclockwise. Illustration on following page shows results.

moves poleward. Colder air with its greater density moving toward the equator forces warmer air with less density over the top where it can move back toward the poles.

Think of cold air moving toward the equator as a large flat bubble spread over a large area wedging its way under warmer air. Edges of these bubbles are called *fronts*—a line at the surface separating warm air from cold air. When a front passes a given point on the earth's surface, the temperature at that point can change by several degrees.

## Motion on a Rotating Platform

The motion of cold-air bubbles is not quite so simple as I have pictured it because the earth rotates. Consequences of motion observed from a rotating platform can probably best be understood by doing a simple experiment.

Suppose you take a rotating turntable (such as a record player), and rotate it from west to east (counterclockwise) as does the earth. (Actually, record players rotate the opposite way, or clockwise.) Instead of a record, use a stiff paper disc. While turning the turntable, move a felt pen across any diameter at a uniform speed.

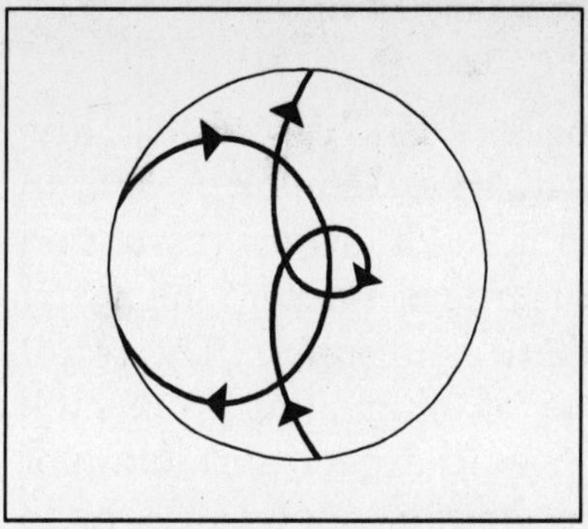

Tracings on disc always go to the right, or clockwise, as do paths of cold-air bubbles over surface of the earth.

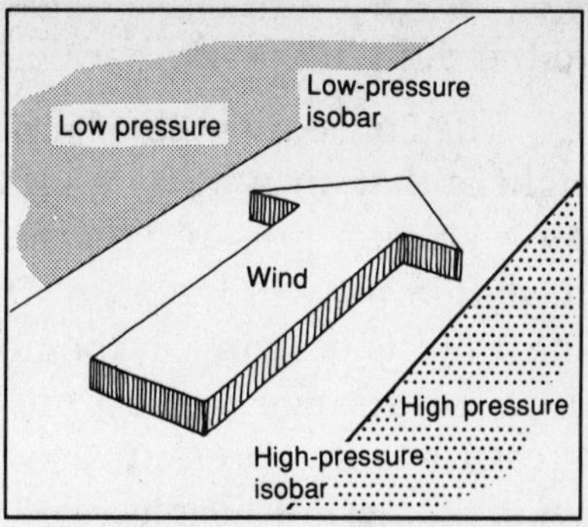


Higher pressure areas always exist to the right of wind movement.

If the record player is at rest, you will get a straight line. If the record player is turning, the trace will be curved. The amount of curving depends on how fast you move the pen and the speed of rotation of the disc.

A common property of all the curves is that they curve to the right. Curving is always to the right whether you move toward the center or away from the center. It makes no difference which diameter is used. The curve on the paper is the motion that would be observed by an observer riding on the rotating disc. We observe all air motions from a rotating earth on which we ride while observing the weather. All air motions are complicated by this curving effect.

## Air Motion on a Rotating Earth

Our simple experiment is similar to the air (pen) moving over the rotating earth (the turntable) in the northern hemisphere. If you were in the southern hemisphere, you would view the paper from the bottom side, and all curvature would be to the left. In the northern hemisphere, the curving is to the right whether you move from the pole toward the equator, from the equator to the pole, from east to west, or from west to east. Any moving air will always be deflected to the right in the northern hemisphere.

The consequences of moving air being deflected to the right is that air will pile up on the right side of its path to give a higher pressure on the right and a lower pressure on the left in *any* wind motion. This is true no matter which way the wind is blowing in the northern hemisphere. The result is rather strange because instead of air moving from high pressure to low pressure as you would expect, air motion or wind is at right angles to a line joining high-pressure to low-pressure regions.

Flowing water also experiences this deflection to the right in the northern hemisphere. Baer, in 1860, noted that the banks of all rivers in the northern hemisphere are steeper on the right side of the flow direction—no matter which way the river flowed. Deflection to the right causes the river to press harder on the right shore, causing more erosion. When he went to the southern hemisphere, he found the same result for left river banks.

Because higher pressure builds up all along the right side of the wind path, think of there being a ridge of higher pressure along a line. Meteorologists calls this ridge of constant higher pressure an *isobar.* Similarly, on the left there will be a line of constant lower pressure to form another isobar.

Pressure will have different constant values along different isobars. A meteorologist will have barometer readings from many different stations so that he or she can construct isobars on a map by connecting points of the same pressure.

## Isobar Maps

After the meteorologist constructs a map from pressure measurements, he or she finds that many of the isobars (lines) form closed paths. If the lines close on themselves, it means there must be centers of low pressure called *lows* and centers of high pressure called *highs.* As you move outward from a low-pressure center, you cross isobars of higher and higher pressure. As you move outward from a high-pressure center, you cross isobars of lower and lower pressure.

The meteorologist can also draw the direction of the wind at each reporting station. He or she will find that winds blow parallel to the isobar curves. By blowing parallel to the isobars, which tend to be closed lines, winds show a giant swirling motion around low-pressure centers in a *counter-clockwise* direction and around high-pressure centers in a *clockwise* direction. This swirling motion always keeps the higher pressure to the *right* of wind motion; lower pressure is to the *left*.

From all of this, you can discover a simple rule known as the *Buys-Ballot law:* Stand with the wind blowing at your back and there will be a low-pressure center somewhere to your left.

## Fronts

I have already described how bubbles of colder air can move along the surface of the earth. How do these bubbles fit into low-pressure and high-pressure systems?

As a bubble of colder polar air encounters warmer air to the south, air with different densities interact. I have described how fluids with different densities can develop waves at the separating surface. In this wave pattern, a crest of warmer air will extend northward while troughs of colder air move southward. As seen from above, the crest tends to become a low-pressure center with a counterclockwise swirling motion about it.

At the surface, there will be two lines extending outward from the low-pressure center that separate warm and cold air. One of these lines is called a *cold front,* the other a *warm front.* Meteorologists designate the cold front with triangles along the line, the warm front with semicircles.

Remember that there is counterclockwise swirling motion so winds are from cold air to warm air at the cold front

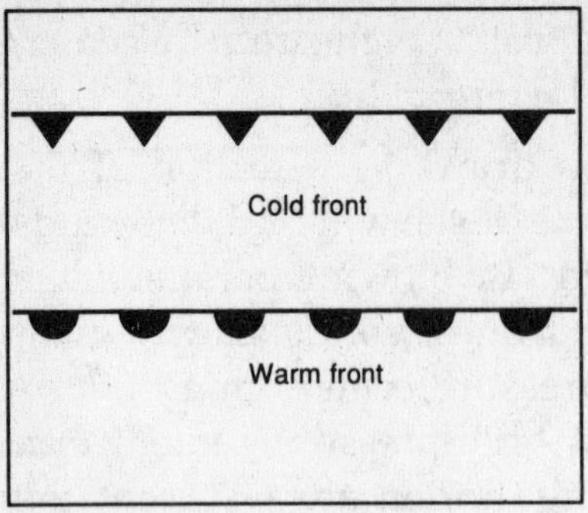


Triangles on a line represent a cold front, semicircles a warm front.

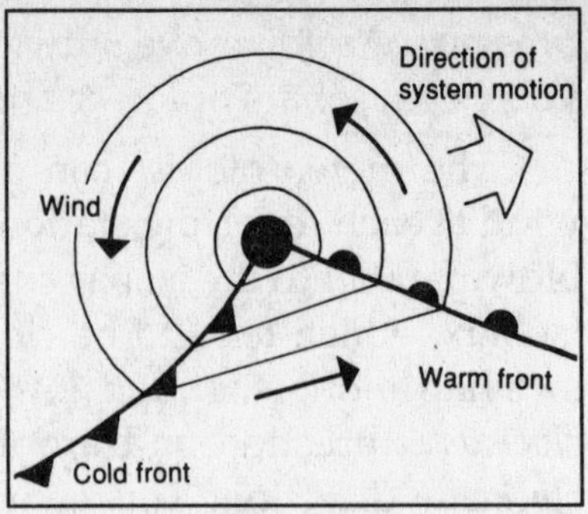


Air flows in a counterclockwise direction around a low center of pressure.

and from warm air to cold air at the warm front. The whole wave pattern then proceeds to move from west to east in a direction approximately parallel to the isobars that extend from the cold front to the warm front.

In isobar diagrams of curved lines surrounding a low-pressure center, there are kinks in the isobars as they cross fronts. Because winds tend to blow parallel to isobars, there will be a rather sharp change in wind direction when a front passes. This wind change is a good indicator that a front has passed.

In the warm-air sector between fronts, winds tend to blow from southwest to northeast. In general, this wind direction gives a clue to the direction that the whole system is moving. In mid-latitudes, this is most often toward the northeast.

You should visualize that along a cold front, cold air is wedging its way under warm air, forcing it upward. If there is moisture in the air, it condenses as rising air cools and leads to thunderstorms. At a warm front, warm air overrides cold air, but again it must rise to do so. Again, there can be lots of precipitation at a warm front.

Cold fronts tend to move faster than warm fronts, creating a scissor action that eventually dissipates the whole system.

Simply by noting wind directions and their changes with time, you can do a pretty good job of describing the entire system and what kinds of changes to expect.

Because fluid-density change is so small at fronts that separate warm and cold air, waves that develop are very long and extend for hundreds to thousands of miles across the continent. When this density change is large, such as going from water to air, wave lengths are of the order of feet and yards. A knowledge of fluid dynamics, which I just touched on, is part of learning the science of meteorology.

# 16 Vertical Motions of the Air

## Convergence, Divergence and Friction

An understanding of *convergence, divergence* and *friction* will lead to an understanding of how horizontal wind motion leads to vertical motion and cloud formation.

In the last chapter, I showed how winds blow parallel to isobars. Near the ground you must consider the forces of friction in addition to pressure forces. Friction changes air flow near the surface.

Close to the earth's surface there are frictional forces that are somewhat greater over land than over sea. These frictional forces make bushes and trees sway in the wind. Above about 500 meters (1640 feet), friction is quite negligible, but from this height downward, friction affects motions of the wind. Friction plays a very important role in the formation of weather.

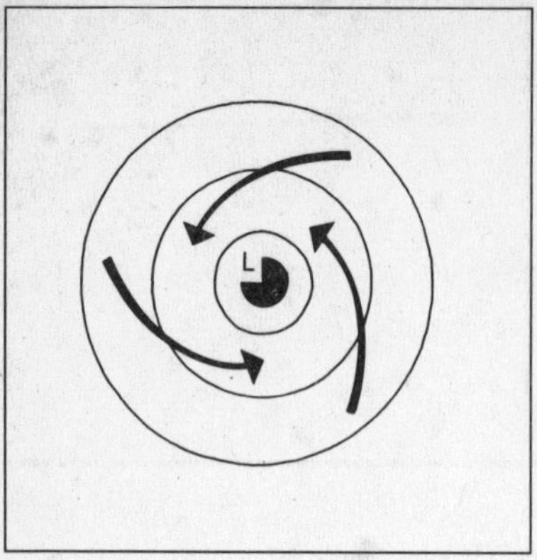


Convergence: Wind spirals counterclockwise into low-pressure center.

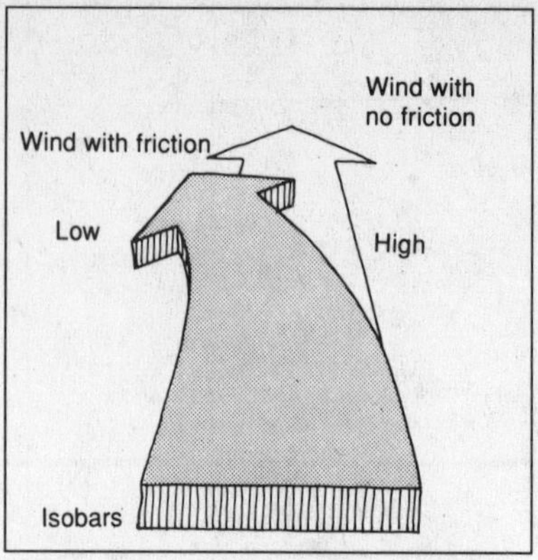


Frictional forces at the earth's surface oppose the wind, slowing its speed and changing its direction.

If you add a frictional force close to the ground, it opposes wind motion and will decrease wind velocity. With a lower velocity, the pressure difference between the isobars is too large and will cause the wind to flow toward the low-pressure isobar. We say that *friction makes the wind cross the low-pressure isobars.*

If this action takes place along a closed isobar, you can see that the winds spiral into the low-pressure center. This happens only near the earth's surface where frictional forces exist. The inward motion toward a low-pressure center is called *convergence.*

A similar thing happens around a high-pressure center, but a low-pressure isobar is always on the outside around a central high. The clockwise-circulating wind will cross the low-pressure isobar so it has a general outward-spiraling motion away from the central high. In this case, we say there is *horizontal divergence.*

Horizontal convergence in a low-pressure region means that air must rise vertically to remove inward-flowing air near the ground. Similarly, there must be downward flow from above to supply the outward flow away from the central high.

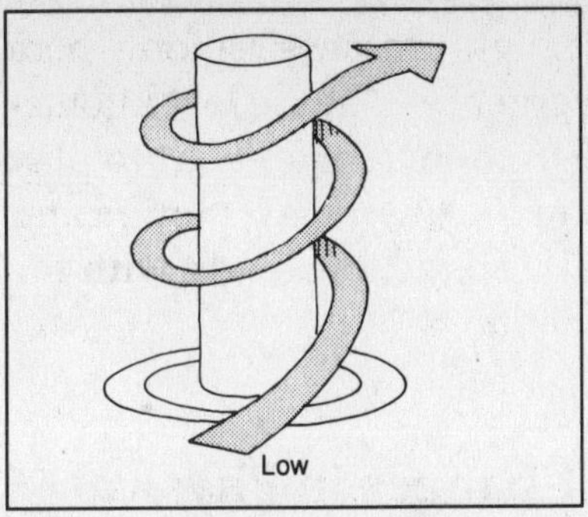


Horizontal convergence in low-pressure region means inflowing air from ground must rise.

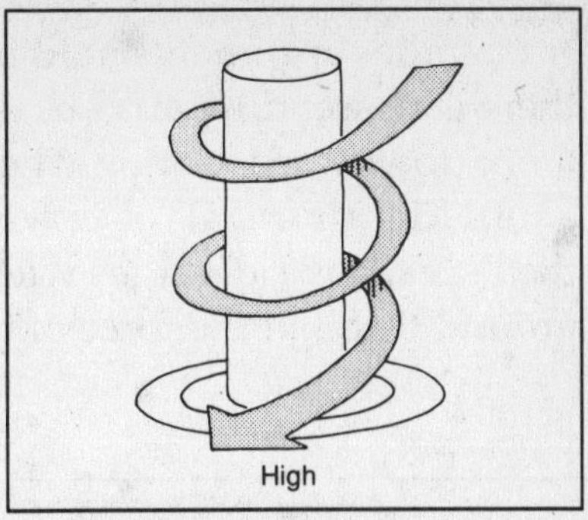


Horizontal divergence requires a downward flow of air from above to supply outward flow from a central high.

Obviously, in the upper regions of the atmosphere, there must be divergence of the air above a low-pressure region and convergence above a high-pressure region. Much of the upper atmosphere motion is realized in interactions with the jet stream. Although upper air motion is very important, we won't consider it in detail.

Most observations of the upper atmosphere can be made only with balloon soundings or satellites. As laymen, most of our observations are restricted to regions near the surface. In some phenomena, like the spreading of anvils above thunderstorm clouds as shown on the next page, you can observe the diverging air from the ground.

Because thunderstorms grow in low-pressure regions, you will be observing convergence near the ground and divergence in the high anvils.

Convergence or divergence near the ground does not extend above one kilometer (3281 feet) in altitude because frictional forces vanish as you move upward. Once upward motion starts in a low-pressure region, frictional forces vanish and upward motion continues as a simple spiral motion. The wind moves parallel to the isobars once friction vanishes.

Similar arguments hold in a high-pressure region where there is descending air. The divergence is in the last kilometer of downward motion where frictional forces start to play a role. The air spirals clockwise in its downward motion and then spreads or diverges when it interacts strongly with the ground through the force of friction.

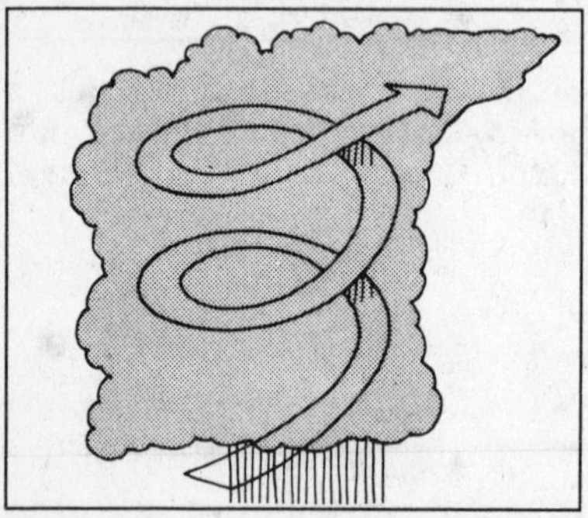

Thunderclouds are characterized by an anvil shape at top of a high cloud where divergence occurs.

## Changes in Physical Quantities

I have explained how air is a gaseous fluid that can be compressed. It is held on earth by the force of gravity. This also means air must support the weight of all the air above it. This weight determines the pressure of the air in that region. I also discussed how air pressure and temperature determine air density. The expansion and compression of air as it moves upward or downward results in work being done. When air expands, its temperature must decrease as it does work on the surroundings. When air is compressed, its temperature must increase as work is done on it. As air rises in a low-pressure region with convergence near the ground, it cools as it rises. As air descends in a high-pressure region, its temperature rises.

## LeChatelier Principle

We know that decreasing the pressure on a mass of air allows it to expand and lowers its density; decreasing temperature makes the air mass contract and increases density. Conversely, increasing pressure makes air contract (compress) and increasing temperature makes air expand. This contrary behavior is the consequence of a very important physical law known as *LeChatelier's Principle.* It says: If any external changes are imposed upon a system, the system will respond in such a way as to oppose the change. In rising atmospheric air, the external change of decreasing pressure allows air to expand. The response of the air is to lower its temperature, making the air contract. That is the response of air to the external change of lowering its pressure is to reduce its temperature and oppose the change.

Conversely, in a high-pressure region where there is descending air, there will be compression, making the volume smaller. According to LeChatelier's Principle, air then gets warmer in an effort to oppose the change. It really makes no difference whether you use LeChatelier's Principle or if you use energy-bookkeeping methods to figure out and remember all of the relations between physical quantities, because either method will tell you what to expect.

Because of the many vertical motions and because air can be heated or cooled in other ways, different regions of the atmosphere at the same pressure can have different temperatures and, hence, different densities. Part of this is due to the air receiving radiation energy from the sun or the air radiating heat to space. Cool air will be more dense than warm air if two air masses are at the same pressure. The atmosphere will have varying amounts of stability because of these effects.

## Temperature Inversions

One of the more stabilizing effects in the atmosphere is *temperature inversion*—warm air above colder air. If the lower cool air then rises for some reason, it will be cooled further and become even denser than the warmer air it was lifted into. Being denser, the cold air can only fall back to the lower starting point. This is then a stable situation in that the temperature inversion stops the vertical motion of the air.

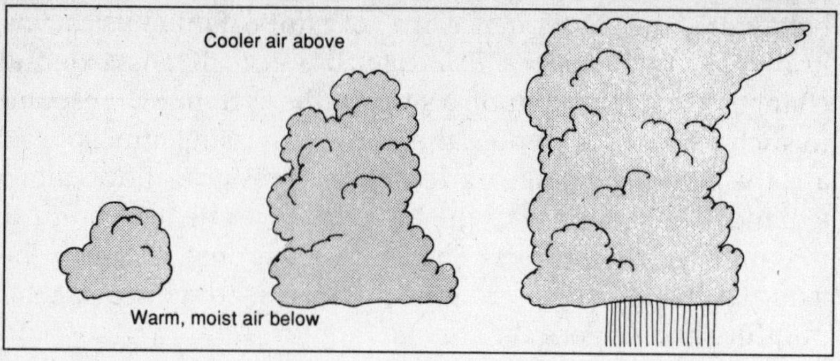

Cloud growth results from rising unstable, warm, moist air.

A common cause of temperature inversion is the downward motion of air in a high-pressure region. This downward motion compresses and heats the air. This may happen over some quite cold air at the surface, so we get a situation in which there is warmer air above colder air. We then have a *stable* temperature inversion.

When this happens over metropolitan areas, the stable situation allows pollution to collect below the inversion. Pollution can only spread at the temperature inversion and fill the cold air with more pollution. This happens during periods of high pressure when air from above descends.

The opposite of a temperature inversion is colder air above warmer air, which is an unstable situation. The warm air below may rise and be cooled due to expansion, but often it is still warmer than the cold air above, so it continues to rise.

A most common cause of instability is when clouds start to form. As moist air rises, it cools and starts to condense water vapor as soon as the air is cooled to the dew-point temperature. When water vapor condenses, it gives up its *heat of condensation* and warms the surrounding air, making it more buoyant. (This is heat that was added at the earth's surface to evaporate the water.) The warm cloud is then a large bubble of warmer air that can be buoyed to higher altitudes and bring in more moist air from below. As water vapor in the additional air condenses, there is more heating and the instability is enhanced. Clouds and precipitation from them usually come in low-pressure regions where rising moist air is normal.

In periods of drought there still may be low-pressure regions, but if the lower air is very dry, there can be no moisture to condense. Cloud bases are very high, and the clouds don't have sufficient vertical depth to develop precipitation. The lower the cloud bases, the more likely rain will occur.

## Water Vapor in the Atmosphere

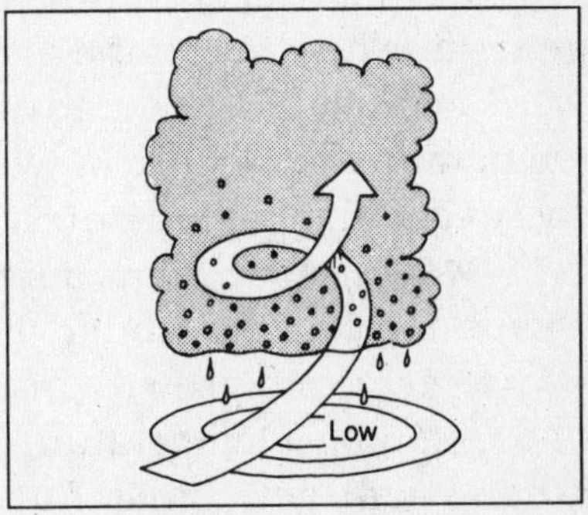


Most water vapor in the atmosphere is at low levels (lower 2 kilometers, or 6562 feet).

All water vapor in the atmosphere evaporated from the surface of the earth, mostly from the oceans. Water vapor then tends to accumulate near the surface, so most of it is found in the lower two kilometers (6562 feet) of the atmosphere. If water vapor is transported to a higher altitude, cooling by expansion makes the temperature decrease to the dew-point temperature so a cloud base forms. At the lower temperatures of a cloud base, water molecules can stick to aerosol particles and form small cloud drops. Eventually some drops grow larger so they can move downward with respect to others and collide with them where the small drops stick to larger ones, growing to precipitation size. This process keep most airborne water rather close to the earth's surface. Mechanisms for forming rain are then all in order.

Swirling counterclockwise motion in a low-pressure center leads to convergence, then rising of the air. If condensation forms, instability is enhanced by warming from the heat of condensation. Low-pressure centers then become the centers where most precipitation occurs. A falling barometer will be a forecaster for rain or snow. I have also discussed how the interaction of warm and cold air at fronts that extend from a low-pressure center can also enhance the lifting of moist air.

In high-pressure regions, descending air brings down dry air from above. The water was removed when this air rose in low-pressure regions. This dry air is not necessarily cold because it was compressed and heated in its downward motion. The high-pressure region will then have clear skies. So a rising barometer means fair weather.

The final temperature of downward-moving air depends much on the temperature at which it started its downward motion. I have already discussed how it may lead to a temperature inversion. In winter, our coldest weather is associated with high pressures. This is because clear skies with little water vapor allow much of the surface radiation to escape and cool the ground. Downward motion is the wrong way to get cloud growth. Instead, there's a high probability of developing a temperature inversion. The cold air near the ground will get colder due to loss of heat by radiation through the clear skies.

You probably have observed that land regions tend to heat more than regions covered by water. Continents warm more than the oceans in the summer and cool more in the winter. Instabilities develop more easily over warm continents in summer. The opposite is true in winter when instabilities can develop more easily over oceans. In winter, continents tend to be covered by high-pressure regions, while in summer there are more low-pressure systems over land.

# 17 Fronts & Winds

## Fronts & Winds Review

*I* have already discussed how warm fronts and cold fronts extend from a low-pressure center and form a dividing line between colder air from the pole and warmer air that comes mostly from the Gulf of Mexico off the coast of the United States.

## Warm Fronts

Let's look at warm fronts first. Recall that we start with the cold air in the form of a large, flat bubble moving in a southward direction. The slope of the surface separating the cold air and overriding warm air is almost flat, rising about 1 mile in a horizontal distance of 100 miles. The sloping surface will then extend over 500 to 1000 miles ahead of the front line at the surface, with cold air below warm air all the way.

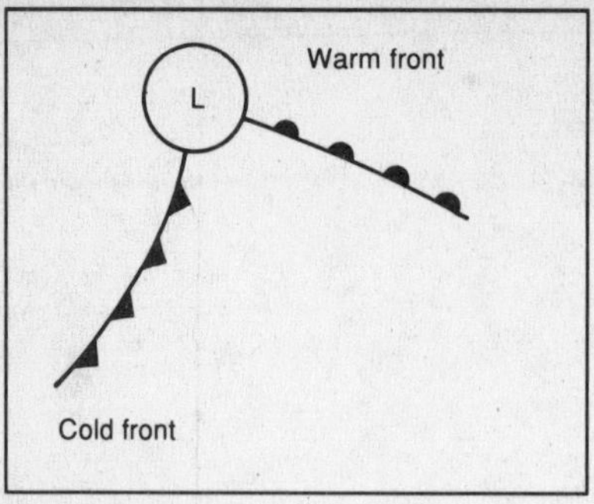


Cold front and warm front extend from low-pressure center, forming a dividing line between air masses such as cold air from the North Pole and warm air from the Gulf of Mexico.

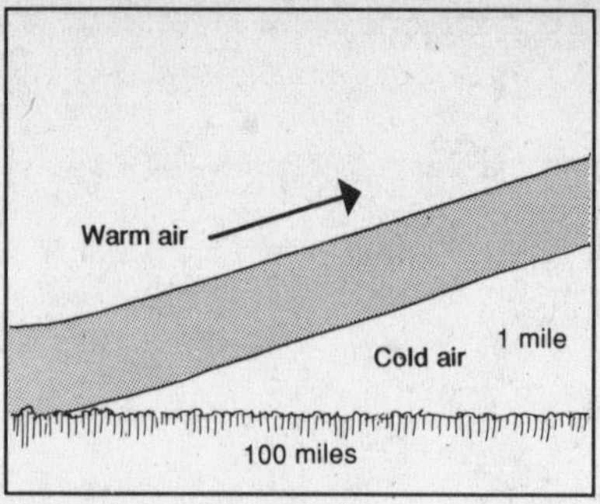


As warm air overrides the edge of the cold air bubble, a temperature inversion is created and cloud formation and condensation start.

The first thing to notice is that there will be a temperature inversion at the surface that separates the two air masses. When a temperature-sounding balloon is sent up in a cold-air region, the indicated temperature decreases until it reaches the separating surface, then takes a sudden rise of a few degrees as it passes through this surface. It then decreases again as the balloon continues to rise. This means that cold air is trapped below warm air, creating a generally stable weather situation. Since the thickness of the cold air bubble can be about 10 meters, this sloping surface will then extend to as far as 1000 miles ahead of the surface of the warm front.

Even though the situation is stable, warm air is being forced upward over colder air by general circulation motions. This overriding warm air can pick up considerable moisture before being forced upward. As the warm air moves up the slope, condensation and cloud formation start. The clouds all remain in the upper warmer air. As air moves up the slope, the clouds get thicker and eventually lead to precipitation. This precipitation can then fall through the cold air below.

Rain can fall for a distance of about 3 kilometers (9843 feet) before it evaporates, leading to a band of precipitation about 300km (186 miles) wide that reaches the ground. The

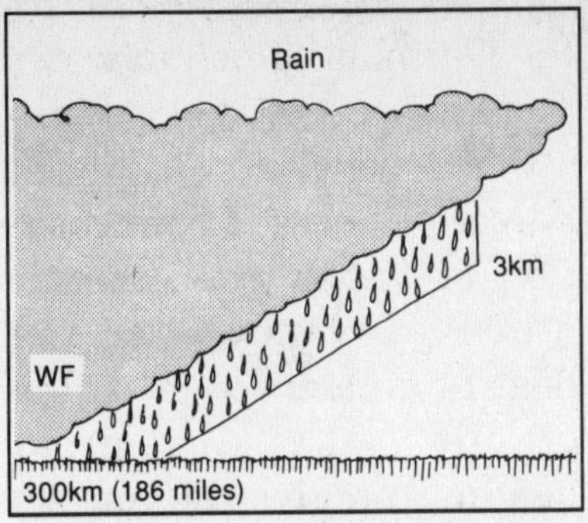

Rain from warm-front clouds falls about 3 kilometers (9843 feet) before it evaporates. A 300km-wide band of rain reaches the ground.

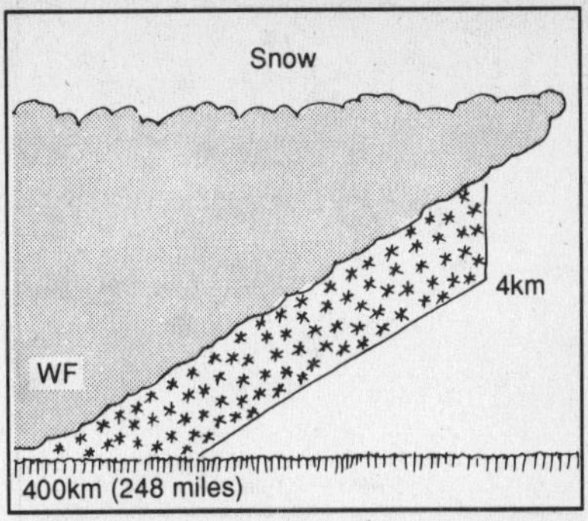

Snow from warm-front clouds falls about 4 kilometers (13,124 feet) before it sublimes. As compared to rain, a 400km-wide band of snow reaches the ground.

front moves about 25km per hour (15 miles per hour), so the warm-front precipitation can create about 12 hours of steady rain.

Snow can fall about 4km (13,124 feet) before it sublimes (goes directly to vapor), so a band of falling snow is about 400km (250 miles) wide and takes about 16 hours to pass. Warm-front weather usually lasts longer in winter than summer. In both cases the weather is warmer after a warm front passes.

Warm-front clouds can extend to a height of 10km (32,736 feet). *Cirrus* clouds (ice crystals forming "mares' tails") will be the first sign of the approaching warm front as they come into view at these high altitudes. The precipitation can still be far over the horizon when these cirrus clouds appear. The clouds gradually thicken to become *alto cumulus* to give "mackerel skies," then thicken further to become precipitating *nimbus* clouds.

## Cold Fronts

The wedge of air at a cold front is considerably steeper and moves almost twice as fast as a warm front. This moving wedge of cold air can push up warm air quite rapidly. When

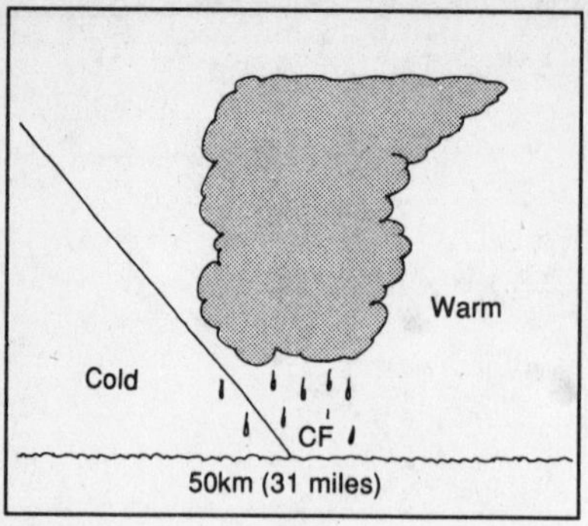

Warm front sheds its moisture as faster-moving cold front forces itself underneath.

the warm air is rich in water vapor, we can get severe storms as the cold front passes. The rapid condensation of vapor gives the air much buoyancy from the heat of condensation so the clouds can rise to great heights and give us quite severe storms, especially in summer when there is much vapor in the air.

With the steeper front, precipitation bands are about 50km (31 miles) wide and will last for only about an hour. After the rain, you are in cold air and wind shifts to the northwest. The temperature can drop by 10F when the front passes and the wind shifts.

You can often see warnings of cold fronts approaching. Fast-rising thunderstorm clouds get sufficiently tall to get caught in the eastward-moving jet stream which blows off their tops and carries ice crystals to the east ahead of the storm, giving the cloud its anvil shape.

You will be in colder air when you see high clouds of a warm front; warmer air when you see high clouds accompanying a cold front. You should then be able to tell which kind of front is approaching. A cold front will usually arrive with much less notice.

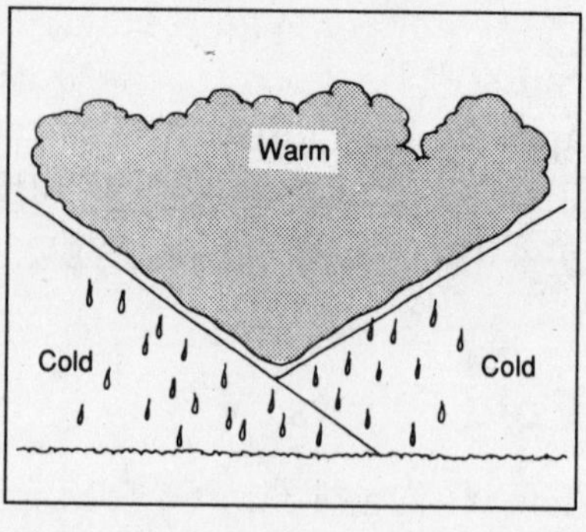

Warm front is "wrung out" when occluded—warm front is trapped between two cold fronts.

## Occlusions

Because a cold front follows a warm front, and because a cold front moves faster than a warm front, a cold front will catch up and pass a warm front. You then have two cold-air masses joining with warm air being pushed up over all of the colder air. This rising warm air is, in general, shedding precipitation, and the sky is very overcast with clouds. Eventually the system wrings itself out and the low-pressure system dissipates.

It is the occlusion that tends to put the head on the comma-shape cloud pictures we get from satellites.

## Changing Winds with Altitude — Veering and Backing Winds

Another aid in weather observation comes from the observations of cloud motions. Clouds at different altitudes may be moving in different directions. You may have a west wind at the ground while clouds at 5km, may be moving from the northwest to the southeast. On other occasions, you may have a west wind at the ground and observe that high clouds are moving from south to north. What can give these changing directions?

The meteorologist always designates wind direction by the direction *from* which the wind is blowing. A north wind means that the wind is *from* the north; a south wind means that the wind is *from* the south, etc.

A second designation tells how the wind is changing direction with altitude. If the wind swings around from the east to the south and then to the west as you move upward, or clockwise as viewed from above, we say the wind is *veering*. If the wind changes from east to northeast to north as you move upward, it is turning counterclockwise when viewed from above, and we say that the wind is *backing*.

As you see clouds moving in different directions, you can determine whether the winds are veering or backing with altitude when there are a few clouds in the sky. As an example, suppose you have a south wind at ground level and higher clouds indicate a more westerly direction. You would say the wind is veering with altitude. If there is wind veering with altitude, warmer air is moving into the region; if there is backing with altitude, colder air is moving in. As a memory aid, use the relationship **W** = **V** and **C** = **B**. **W**arming accompanying **V**eering = letters at the end of the alphabet; **C**ooling with **B**acking = letters at the beginning.

The explanation for these complementary relationships between wind direction and air-mass movement is a little difficult. Start by supposing there is a vertical wall of cold air next to a vertical wall of warm air. Density of the colder air is greater so the pressure decreases more rapidly as you go upward in the colder air. If the pressure at the ground is the same for both columns, then as you go upward, pressure will always be lower in the colder air at any given height. If you could draw in isobars at any level, the high-pressure isobar would be in the warmer air. You can then a pply the Buys-Ballot law to find that the wind blows parallel to isobars with low pressure on the left when the wind is at your back. This is called a *thermal* wind because it arises from temperature differences in the air masses.

A rule for finding the thermal wind is: With the thermal wind at your back, the low temperature will be to your left.

As you go upward in the atmosphere, a thermal wind can change wind direction. Suppose that you start with a south wind at the ground and as you go up the cloud motion shows that there is veering around to the west. As you go upward, you will be adding a component of westerly wind due to the thermal wind. With a thermal wind from the west at your back, the low temperature will be to the north and the higher temperature to the south. The southerly wind at the ground will then bring in

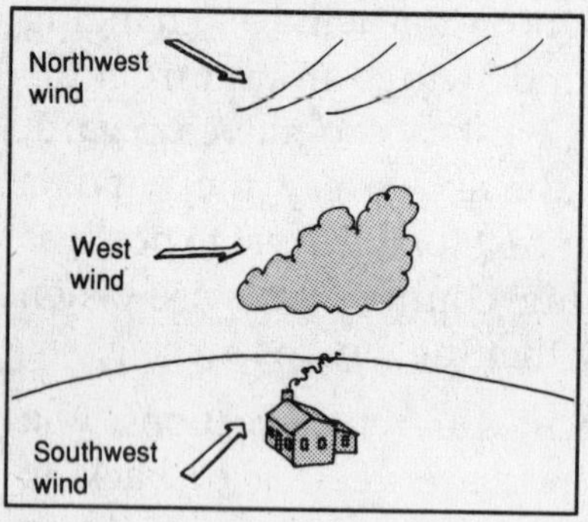

To determine whether the winds are veering or backing with altitude, look up at the clouds that are at different levels to see what directions they are moving.

warmer air from the south so that veering goes with warming. If there had been backing, the thermal wind would have been reversed and warmer air would have to be to the north. The original southerly wind at the ground would then bring in colder air so that backing goes with cooling.

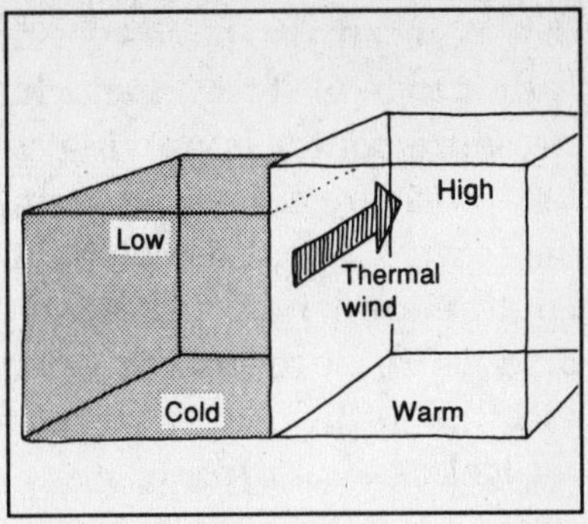

To find thermal wind, stand so wind is at your back; low temperature will be to your left.

## Overall Circulation of the Atmosphere

You may wonder why our northern hemisphere weather system always moves from west to east. We can apply our thermal-wind concepts to overall circulation. In general, there is always colder air to the north and warmer air to the south. If you stand with the low temperature at your left, the thermal wind at your back will be from the west. This thermal wind carries our weather systems from west to east.

Sometimes when you look upward you see both veering and backing at different levels. At some levels, it may be getting colder, and at other levels it may be getting warmer. In my discussion of vertical motion, you read that these upper air temperatures greatly affect the stability of the air. If you observe veering in a region above, you will have warmer air moving in above and a more stable situation developing. So you can predict fair weather. If you observe backing wind in a region above, you will know that the region is filling with cooler air above to make the situation more unstable and you can predict foul weather. If you find backing above and the air is not humid at the surface, expect severe storms and possibly tornadoes.

You can also have veering or backing of winds with time rather than altitude. Such changes can indicate where a low-pressure center is moving with respect to your area. If the low-pressure center is passing to your south, don't expect to encounter a passing warm front, then a cold front. But you could experience weather typical of that for an occlusion. If a low-pressure center passes to your north, expect first weather typical of a warm front and then weather that is typical of a cold front.

# 18 Why the Sky is Blue

## How Light Interacts with Matter

Sunlight you see as white light is a collection of many different colors. If a beam of sunlight travels through a small opening and then through a glass prism, it breaks into many colors we call a *spectrum.* All colors are bent (refracted) by the prism, with blue light being bent the most.

A physicist differentiates between colors by knowing each one has a different form of wave motion. Red light has a long wavelength; blue light has a relatively short wavelength. Because blue light is bent most by the prism, we say the color with a short wavelength interacts *more strongly* with the prism.

Blue light from the sun also interacts more strongly with the atmosphere and particles in the atmosphere as it passes through. This stronger interaction is why the sky is blue. The sky scatters blue light more than red light. The blue sky is

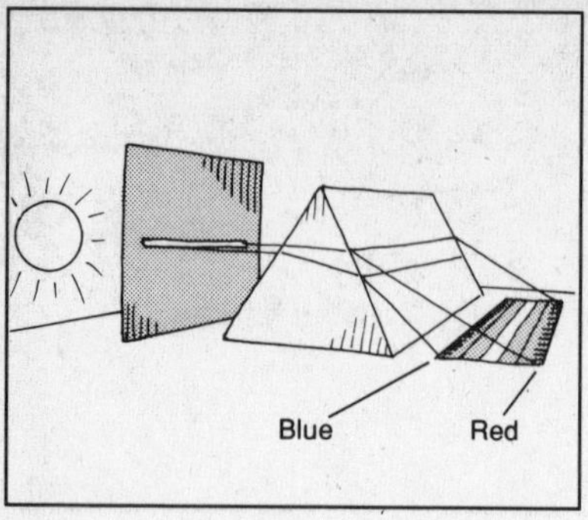


Beam of sunlight passing through narrow slit and then a glass prism is broken into many colors, or spectrum. Red light has long wavelength, blue a short wavelength.

much like shining a strong flashlight beam through dusty air in a dark room where you can see the beam. Particles in the beam scatter some of the light from the beam. If there's enough dust, this scattered light can partially illuminate the room. If the dust particles are sufficiently small, the room will take on a bluish hue because of the preferential scattering of blue light.

Selective scattering of blue light in the atmosphere becomes more apparent at sunrise or sunset when the rays from the sun pass through more air before they reach us. The sun actually appears to be quite red at these times, indicating that essentially all of the blue light has been scattered out of the beam, leaving red light.

Because blue light has a shorter wavelength, the atmosphere appears to be more rough to blue light than to red light. This is just as a sandy beach may appear rough to the small ant, but smooth to the larger human. Blue light interacts with individual molecules or air just as the ant interacts more with individual grains of sand.

This selective scattering can tell us things about the size of particles in the atmosphere. In clear air, the molecules of air and very small dust particles make the atmosphere appear to be relatively rough to blue light and smooth to red light. If water vapor starts to condense on the dust particles, they become larger and red light interacts with them more strongly. The sky and clouds in the sky then become illuminated by red light as well as blue light. The sky takes on a more reddish hue. A sky turning red can often tell us that water-vapor content in the atmosphere has increased.

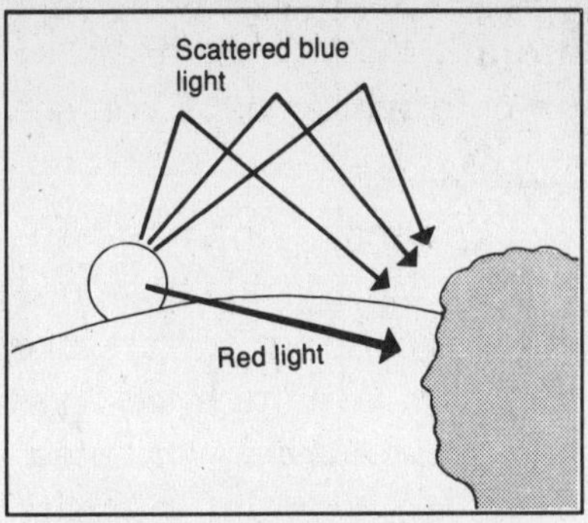


Selective scattering of blue light in atmosphere is more apparent at sunrise or sunset as sunrays pass through atmosphere at a steep angle, leaving red light.

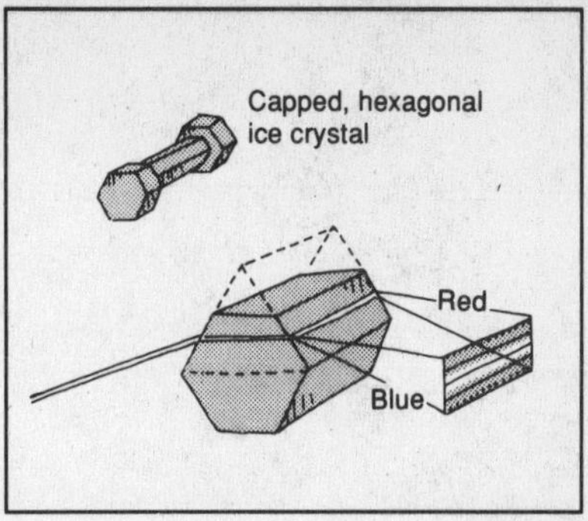


Hexagonal crystal acting as a prism.

## Cloud Illumination

Small water drops in clouds are large enough to scatter both red and blue light so they appear white against the blue sky, where there is mostly blue-light scattering. When dust particles in the air are being made larger by condensing water vapor, the sky can turn from blue to a hazy white. At sunrise or sunset, enhanced scattering occurs with a predominantly red beam of light so the sky takes on a more reddish hue.

A red sky can mean there's a lot of dust in the air, but more often it means that condensation onto the dust particles is occurring. This indicates there is increased moisture in the air.

## Evidence for Water in the Air

When haze in the sky is high, there is condensation at high altitudes, usually in the form of ice crystals being blown into the region by high-altitude winds. These come from high clouds in a front to your west.

Ice crystals have a definite hexagonal shape. (This is why snow flakes have six points.) Sunlight or moonlight can pass through these crystals just as light passes through a prism and bends light in preferred directions. You then see a ring

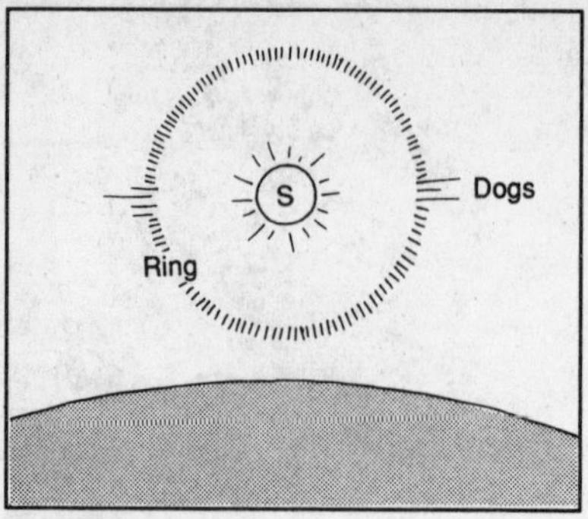


Ring around moon or sun is caused by light passing through ice crystals. On cold winter days, we see sun dogs.

around the sun or moon, or on cold winter days you see *sun dogs.*

You may think it is difficult for ice crystals to be oriented correctly to observe this ring. Although ice crystals are randomly oriented, there is maximum intensity when light passes through them in a symmetrical way. This maximum intensity is observed as a ring. In the case of sun dogs, the ring is incomplete. The observed ring will be between the sun and the observer, and the ice crystals seen by each individual are a different set of crystals.

## The Rainbow

Rainbows are similar to rings around the sun or moon, but differ in that they occur with spherical water drops and there is internal reflection in the drop. The added reflection means a rainbow is always on the opposite side of the observer from the sun. Your back must be to the sun to see a rainbow. This makes it impossible to see a rainbow at noon. You can only see part of the circle of the rainbow when the sun is lower in the sky. If there are raindrops falling on the side opposite you from the sun, you'll see a rainbow if there are no intervening clouds.

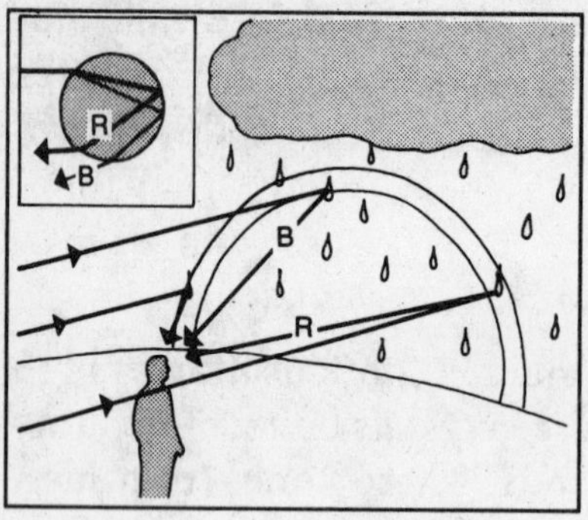


Caused by sunlight reflecting from inside spherical water drops, rainbows always occur on side of observer that's opposite the sun.

## Polarization of Light

When light is scattered, it can take on another property—that of becoming *polarized.* This makes it possible for Polaroid sunglasses to remove glare from scattered light.

Light is an electromagnetic vibration with vibrations *normal* (at 90 degrees) to the direction of propagation. When light interacts with electrons in atoms, it shakes them parallel to vibrations in the wave. The vibrating electrons then reradiate light in preferred directions. The electrons can radiate at right angles to the vibration, but not in the direction of the vibrations. Reradiated light is thus polarized in the direction of electron vibrations.

Experiments with insects, especially bees, show they can detect this polarization and obtain a sense of direction with respect to the sun. A bee finds its way back to its hive by detecting the polarization of scattered sky light. If anything happens to destroy this polarization, the bee loses his sense of direction and will stay close to the hive. Cloudiness destroys polarization, so honey production is poor during cloudy summers.

## Atmospheric Turbulence

Turbulent air high in the atmosphere tends to make light travel in slightly wiggly paths through the air because of variations in air density. These wiggly paths can, for an instant, not let starlight into the eye, so a star will twinkle. Planets don't twinkle because you see them as finite in size. Even though stars are larger, they are so far away that they appear as point sources of light.

When stars twinkle, the air is unstable and turbulent. Twinkling of dim stars actually removes them from view. That's why there seem to be far fewer stars when turbulence develops.

Stars near the horizon actually twinkle with color changes. In the mid-latitudes, western stars twinkle red-green-blue; eastern stars twinkle blue-green-red. Blue light is bent more in the atmosphere so that irregularities in turbulent air cut off blue light first as the air moves from west to east. The reverse happens for eastern stars. When this viewing is done closer to the equator where there are prevailing easterly winds, the order of colored twinkling is reversed.

The new moon has points or *horns.* When the air becomes turbulent, the twinkling effect makes the horns look duller. In stable air the dark part of the new moon is visible because it is illuminated by sunlight reflected from earth. As turbulence develops, the moon's dark side is so dim that it can no longer be seen.

## Seeing Farther Before Rain

There are many proverbs about being able to see better before a rain. While the sky is still clear and humidity starts to increase, aerosol particles in the air can start to become hydrated and grow larger. After growing a little, they become sufficiently large to scatter blue light, so you start to see a blue haze.

The amount of scattering depends on the number of drops per unit volume and the cross-sectional area of each drop. If some drops become a little larger than others, surface-tension forces are less for the larger radius and condensation can occur on these more readily than on smaller droplets. Actually, smaller drops can start squeezing themselves to death with surface-tension forces so that all of the condensed water is transferred to the larger droplets. There are far fewer of the larger drops, but they all have a larger cross section for light scattering. It turns out that this process will decrease light scattering and clear the air because the effect of having fewer drops is greater than having a larger cross-sectional area. You then can start to see more distant objects.

You may wonder how water can condense in particles below a cloud base. It turns out that many of the aerosol particles are slightly soluble in water, and this solution of aerosol particles in water will have a lower vapor pressure than pure water. These small drops of solution can then exist below cloud base. There is other evidence that the aerosols become hydrated in that the sense of smell of animals and man is enhanced when water-covered particles can more readily attach themselves to nerve endings in the nose. Measurements of the mobility of ions also show that they become more massive with hydration when humidity starts to increase.

## Green Sky Before Hail and Tornadoes

Many people notice greenish clouds before severe storms. Most often it is associated with hail, and hail is an integral part of a tornado-bearing cloud.

I can only speculate as to why the sky turns slightly green. These storm clouds are very tall, dense clouds, so that absolutely no direct sunlight gets through. Everything gets very dark, but there must be some light because you can still see. I think that most of this dim light is reflected from green foliage on the earth's surface to the cloud. This is much like the dark part of the moon which is slightly illuminated by reflected sunlight from the earth during a new moon. Plants absorb red and blue light, and reflect any green light. When the light is dim, they also orient their leaves and chloroplasts in the leaves to reflect additional green light. Reflected green light becomes the primary source of illumination for seeing the sky with its dark clouds.

## Hearing Better Before Rain

We seem to hear distant sounds better before rain. This is difficult to explain. Two things change when moisture in the air increases. First, the velocity of sound increases because water molecules are a little less massive than air molecules. Second, high frequencies in sound waves are absorbed more

This second effect is noticeable when you hear thunder. Close thunder carries more of the high frequencies before they are absorbed, giving thunder more of a *loud-crack* property. Distant thunder is always heard as a low-frequency rumble because the high frequencies have been absorbed in the longer path between the source and the observer.

A similar effect can be observed on very cold winter days when jet aircraft fly overhead. The loud cracking noise is much more irritating because the noise includes many more high-frequency sounds. In this case, the extreme cold has condensed out most of the water vapor so very little is left to absorb sound. Farmers can hear better what is going on at a distant neighbor's home. When there's little wind, they can hear their neighbor's cattle mooing or someone chopping wood on cold winter nights.

Before a rain, moisture in the air increases so the velocity of sound is somewhat greater because lighter water molecules replace heavier air molecules, making the air *less dense.* Some people think moist air is heavier than dry air, but this is not true. The ratio of water vapor mass to air mass is the same all the way between ground and cloud base, so the increase in sound velocity from moisture is the same everywhere. In general, temperature decreases as one goes up so sound waves at higher elevations travel slower. Thus sound waves will be refracted upward and make it more difficult to hear distant sounds.

When a warm front approaches, there's a temperature inversion at the cloud base where warm air rides over colder air. This temperature increase makes sound waves travel faster and refracts them back toward the earth. On cool still evenings at a lake, a temperature inversion occurs over the water and one can hear sounds across the lake much better.

In ancient times, distant hearing was usually judged while listening to church bells. These bells were always placed in high towers so the refraction of sound waves at ground level could carry sound a greater distance along the ground surface. A distant church bell could be heard better before rain.

# Lightning

*His lightnings enlightened the world;*
*The earth saw and trembled.*
*Psalm 94:4*

# 19 Lightning

## Lightning Probabilities

What are the odds of being struck by lightning? From the frequency of deaths per year in the United States due to lightning, we can make an estimate for the probability of lightning dealing any one of us a fatal blow. There are around 200 deaths per year in the U.S. population of somewhat over 200 million people. From these two values, we can estimate that the probability of being struck by lightning is about one in a million. Up to now, we have said that this is such a small value that one need not worry. Since the introduction of lottery games, we find many people buying tickets and hoping to win when the chances are one in 12 million. If we are willing to accept such odds, we should perhaps worry more about lightning.

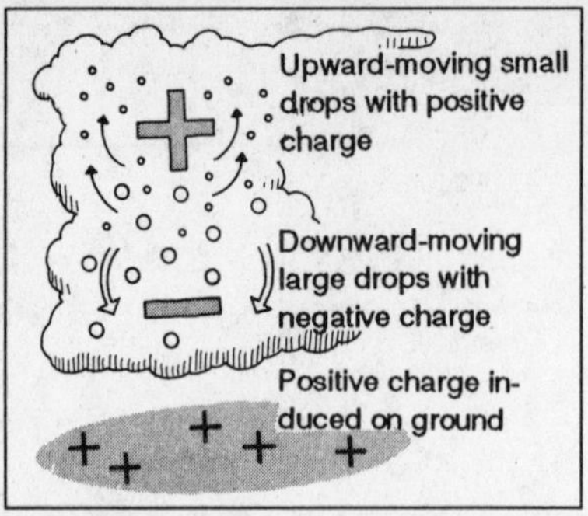


A thundercloud is positively charged at its top. Negative charges at its base can induce a layer of positive charge on the earth directly underneath, establishing a conditon where lightning can occur. There are both large and small hydrometeors in the middle of the cloud.

## Charging in a Thunderstorm

Before listing ways for changing the odds, I'll describe how lightning discharges occur. Lightning is so capricious that it is difficult to write a sufficiently complete set of safety rules to cover all situations. For this reason, let's look at some of the scientific details of how the charges are generated and how these charges are collected into intense lightning currents. These currents can do damage in many different ways as they bring charges from the clouds to ground. An understanding of these processes can be an excellent guide to how one can cope with the strange things that lightning can do.

In a thunderstorm cloud, large drops of water and ice falling toward earth give us precipitation. These large *hydrometeors* encounter many smaller, lighter hydrometeors being carried upward by the thunderstorm updrafts. When they meet in midair, the small droplets lose some of their water to become smaller and make the large drops larger. Along with the transfer of water between the drops at different temperatures, there is also a transfer of negative electric charge in the form of electrons from the smaller drops to the larger drops. This, in turn, leaves the smaller upward-moving drops with a deficiency of negative electrons and hence makes them have a positive charge.

As the process proceeds, the cloud becomes positively charged at its top and negatively charged at its bottom. The negative charge at the base of the cloud then can induce a layer of positive charge over the conducting earth in the region below the cloud. When you stand beneath a thunderstorm cloud, the negative charge at the base of the cloud repels many electrons from your body into the earth. This leaves you covered with a layer of positive charge which tends to concentrate on the highest portion of your head. In some cases, the hair on your head will rise and point toward the negative charge in the cloud. Most of the lightning strokes to ground are then discharges between the negative charge in the base of the cloud and the positive charge induced on the earth's surface. If you somehow get in the way of the discharge, you are in trouble. There are about four times as many discharges between the negative cloud base and the positive cloud top, but they do no harm to objects on the ground.

## Ions and Free Electrons

When we take a more microscopic look at things, we find them to be made of *atoms* or *molecules*. The usual state of these molecules is to have a positive nucleus surrounded by negative electrons so that the overall system is electrically neutral. If a negative electron is removed from a molecule, the molecule has a net positive charge and is called a *positive ion*. Usually the removed electron quickly attaches to a neutral molecule to make it become a *negative ion*. It is this transfer of electrons that allows the cloud to become charged with an excess of negative ions near the cloud base and attached to rain drops. Similarly, this electron transfer allows an excess of positive ions attached to small particles in the upper regions of the cloud.

While the electrons are moving between molecules to form a pair of oppositely charged ions, we have a *free electron*. It is the motion of free electrons that forms the lightning current. Free electrons are at least 2,000 times lighter than molecules so that if strong electric forces are present, the free electrons are accelerated to high speed and can acquire sufficient energy to free more electrons from neutral electrons during collisions with them. This creates an avalanche of free

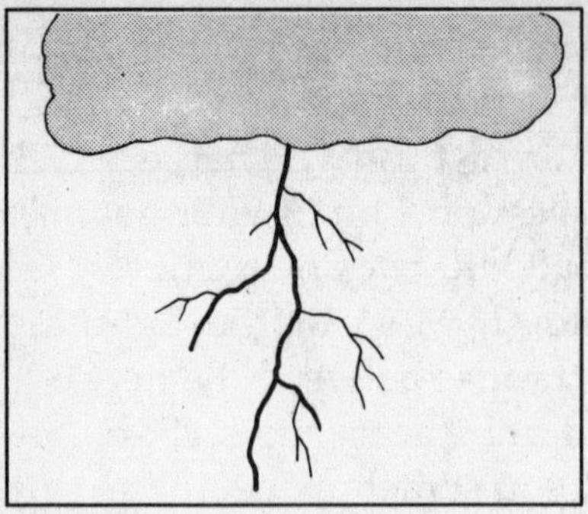

First lightning stroke is a branched light channel, usually directed downward . . .

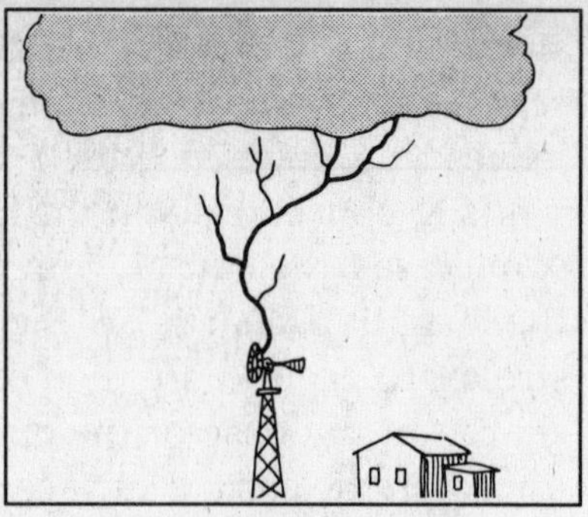

. . . but is sometimes directed upward from a tall object such as a building or tower.

electrons which becomes the lightning current. Lightning processes are so fast that only the free electrons can move an appreciable distance during a lightning stroke. And it is these millions of fast-moving electrons that make lightning so devastating.

Most lightning discharges start near the base of the cloud where there is a large concentration of negative ions on rain drops and a few free electrons. These free electrons, under the influence of strong electric forces, start an avalanching process directed toward ground begun by a *leader stroke* with currents of 200 to 500 amperes. The current follows a very narrow channel because the downward avalanching of electrons where the discharge starts forms a conducting needle-like point in the surrounding non-conducting air. More electrons from the cloud base move into this conducting region to enhance the electric field at the tip.

This is a positive-feedback situation in that the longer the needle-like conducting point becomes, the stronger the electric field at the tip of the channel. The availability of an avalanching electron in the strong electric field at the tip may be slightly to one side of the tip center so that the channel becomes a zig-zag path as the breakdown proceeds in a general direction toward the earth. Sometimes there may be two electrons that can start two new avalanches near the tip so that the channel then forms a new branch.

One of the branches will find its way to ground to complete the lightning channel while the other branch becomes exhausted before reaching ground and forms what appears to be a tributary to the main channel. Sometimes during a close lightning strike you can hear a sort of ripping sound prior to the main crack of thunder. This happens when one of the branches is closer to you than is the main channel to ground. If you hear this ripping noise, you should consider yourself to be very lucky as the main part of the stroke has proceeded to some other point at ground.

It turns out that it is easier to start a positive discharge where electrons are sucked onto an electrode than it is to start a negative discharge where free electrons are repelled from the electrode. As the leader stroke approaches the ground, the electric forces at ground level become larger and larger and induce greater and greater amounts of positive charge on objects on the ground.

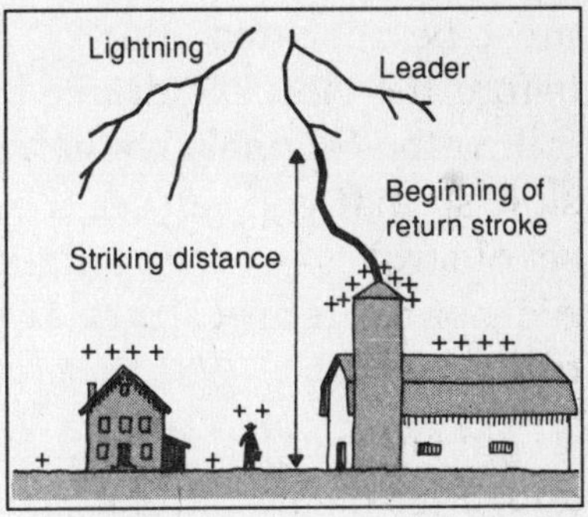


While all the objects pictured here are potential targets, the taller silo has a more concentrated positive charge and is more likely to be struck by lightning.

Tall objects at ground level have larger concentrations at their tops than short objects and eventually the tallest object will start sucking in free electrons to start an upward avalanching process which meets the downward leader stroke about 50 meters (164 feet) above the starting point. This distance is called *striking distance*. This upward motion is the beginning of the return stroke where currents can increase to as much as 100,000 amperes. The return stroke moves all the way to the cloud base to remove negative charges deposited along the channel by the leader stroke with a speed of about one-third the speed of light. It is in this short period with large currents that damage is done.

It should now be more clear why lightning tends to strike tall objects which have tops closer to the downward-moving leader. Thus tall objects concentrate more positive charge on their top regions as the leader moves downward.

The proverb which says lightning never strikes the same place twice is absolutely wrong. In the past, the proverb seemed true because the first time that lightning struck a tempting tall wooden target such as a church spire or a ship mast, the object would be destroyed. A well-grounded metal structure such as a radio transmitting tower will survive many strikes.

## The Complete Lightning Flash

The lightning process previously described is called a *stroke* of lightning. Usually one stroke will not drain all of the negative charge from the base of the cloud. So the stroke must be repeated four or five times in rapid succession before the cloud is drained. All of the subsequent strokes follow the original channel, but the repetition is sufficiently slow so that you can see the lightning flicker. The complete set of strokes is called a lightning *flash.* The destructive currents of the return stroke are repeated in each of the strokes during a flash.

In some cases, the entire flash does not drain the cloud and, after the last stroke in the flash, a continuing current of a few hundred amperes may persist for almost a second or so. This continuing current starts most lightning fires. Bricks being blown from a chimney or other physical damage from lightning is caused by the intense electric and magnetic forces around the stroke. This is called *cold* lightning. If there is a continuing current and a fire, it is called *hot* lightning. The U. S. Forest Service has sophisticated instruments which distinguish between hot lightning and cold lightning. When a hot lightning stroke is observed, they send out crews to the strike point to extinguish the fire.

## Distribution of Lightning

The incidence of lightning varies widely over regions of the earth and with the seasons. At any one instant, there are about 1,200 thunderstorms in progress each delivering about three lightning flashes per minute. A family of satellites photographing the earth at all points on the earth would register about 60 flashes per second. Strangely, the satellites observe that most of the flashes occur over land with very few over the oceans.

This fact was known by the old sea captains who recognized they were near land when they started to see lightning. If we exclude polar regions where there is no lightening, we find that there is about one flash per square mile per year. Because most of us see many more flashes than this each year, it means we can see lightning over an area much larger than one square mile—more like 100 square miles.

## Description of Lightning

Any one lightning flash consists of several strokes separated by about four one-hundredths of a second, making the flash appear to flicker. Some people with fast reaction times can actually count the number of strokes in a flash, which may vary from one to over 20 strokes. A fluorescent lamp flickers every one-hundredth of a second. So, if you can discern the flicker in a fluorescent lamp, you should be able to discern the flicker in lightning. Watch for this flicker and try to count the number if you can.

The duration of a multistroke flash varies from about two tenths to one half second. It takes about one second to say "one thousand one." The average number of flickers is about four. The first stroke is a branched light channel. The branches most often are directed downward, but on a few occasions are directed upward.

Branching is formed by the first leader stroke as it probes its way through the un-ionized air. The first leader stroke usually starts at the cloud end of the channel so that the branches point downward.

Lightning branches are formed as the downward-moving charge from the cloud feels its way toward earth and successfully or unsuccessfully probes its way through the intervening air. However, some leaders start at the top of tall buildings or towers, in which case branching is upward. In this case the charge is induced on the ground and is released from the tall grounded structure to feel its way to the opposite charge in the cloud. Most flashes, however, are of the downward type.

The illuminated sky then appears as a river on a map. The original charge, however, has moved up the river as it forms branches. Called the *leader stroke,* it deposits a charge all along its channel and branches. When one branch gets close to earth, the electric fields are very strong so that when the leader is about 50 meters (164 feet) from ground, a return stroke is initiated, usually from some taller object. This sends a charge of opposite polarity back along its channel to neutralize the charge placed there by the leader. Called a *return stroke,* this happens very fast.

The leader and return make up one complete stroke in the flash. After four one-hundredths of a second, a second leader moves rapidly toward the ground as a *dart leader,* not having to reestablish the branches. When it gets close to the ground a second return stroke occurs. This process is repeated until the flash is completed and may happen as many as 20 times during a large flash.

There will be no repeated flicker in the branches because once a channel has been established, all subsequent strokes can follow the first channel to ground. If you can discern flickers in the main channel, see if you can discern the lack of flicker in the branches.

## St. Elmo's Fire

St. Elmo's Fire: Electrical field too small to propagate into lightning discharge is characterized by hissing sound between discharges atop points such as towers or trees.

Many have observed St. Elmo's fire from sharper points during the time between discharges. This is a case of *corona discharge* due to very strong electric fields in a region that have been enhanced by a point. You should be extremely cautious when this is observed because corona discharges are small discharges that aren't quite large enough to evolve into a propagating lightning discharge.

During the day St. Elmo's fire is often too weak to be seen, but you can often hear it hissing. If you observe the top leaves of a tree when this occurs, you'll see the leaves dancing wildly under the influence of electrical forces. Try to improve your protection when you encounter St. Elmo's fire.

## Ball Lightning

A final electrical phenomenon to watch for is ball lightning. Only a few people have seen this, and there are no good photographic records.

A ball of plasma or fire varies from grapefruit size up to 30 inches in diameter and will last for several seconds after a flash. The ball bounces around and can pass through windows and doors. It usually vanishes with a bang when it encounters a metal stove or some well-grounded object, giving observers a most frightening experience.

Ball lightning is of considerable scientific interest because no one has been able to produce it in a laboratory. There is no good theoretical explanation of how it can live as long as it does. If you observe the phenomenon, you should, in the excitement, try to estimate the size, motion, colors, any smells, its duration and how it vanishes. Try to estimate any damage it may have done. Experts will be *very* interested in your report and would be most grateful to get a picture record.

## Lightning and History

Lightning has played an important part in the lives of people throughout history. It is amazing how thunderstorms in general can affect history through the thoughts of individuals. Earlier, I gave the proverb, "The minds of men in the weather share, dark or serene as it's foul or fair." I then described how the Roman Senate used lightning to tell them when to vote.

As described above, the fear of lightning can change the course of events. There are many interesting stories about how severe storms altered the course of history:

Louis XIII of France was a typical king in his single-minded desire for an heir. He married Anne of Austria, and in due course, she became pregnant as hoped. However, Anne was playful and liked to slide down the banister of the castle stairs. During one of her romps, she fell and had a miscarriage as a result. This made Louis so angry that he would not touch her for the next 18 years.

Louis was an ardent hunter and often hunted around the country castle where Anne was living. On one hunt, after this long time apart, Louis was near her castle when a very severe thunderstorm occurred. He took refuge from the storm in Anne's nearby castle and spent the night in Anne's bed. As a result of this union, she became pregnant again and bore a son who became the Sun King, Louis XIV, who did much to change the history of Europe.

In earlier times, lightning and thunder were usually understood as some communication from heaven. At the end of the thirteenth century and the beginning of the fourteenth, there were many wars between the Municipal States in Italy. The Papal State was so badly defeated that the papacy was moved to Avignon in France in what was called the *Babylonian Exile.* As a result of this move, many Frenchmen became cardinals and France received many benefits from this arrangement. England, Spain, Germany, and other countries objected to having their tithes to the church being turned over to the causes of France. They demanded that the papacy return to Rome. This move had to be voted on by the School of Cardinals who were mostly Frenchmen, so there seemed little chance that a move back to Rome would be realized.

While the cardinals were voting, however, lightning struck the electoral chamber and set it on fire. Thinking this was a message from God that they had better do the correct thing, the cardinals voted to move the papacy back to Rome where Gregory XI became the pope. Later, some of the cardinals had second thoughts and elected Clement VII to remain as pope in Avignon, thereby creating a schism in the Catholic Church. Clement VII and Benedict VIII ruled as *anti-popes* in Avignon during the 70 years that the schism lasted.

During the rule in England by the Plantagenets (1154-1458), England also ruled much of France. (All kings up through George III were always crowned King of England and France.) Edward III and his son, The Black Prince, spent much of their time warring and plundering in France. They even took King John of France back to England as a captive. Both nations were Catholic at this time so the pope made an effort to call the two nations together to work out some peaceful settlement. Edward III demanded that he be crowned in Paris as King of France, but the French insisted that John was still their king and this could not be changed. While the negotiations were proceeding, lightning struck the meeting hall. It was quickly decided that Edward and his army should leave France and return John to be the rightful king of France.

Another example of a thunderstorm changing the course of history is Will Durant's story of Martin Luther, who was originally trained to be a barrister. One night Martin Luther was caught in a severe thunderstorm. In his fear, he promised God that if he would be spared, he would devote the rest of his life to work in the church. He lived and, consequently, retrained himself to be a clergyman. Of course, during this training, Luther found grounds for disagreeing with his peers and the church authority. The Reformation and other major changes in society followed; wars were fought and kings were dethroned as European countries engaged in the Thirty Year War, which was fought mainly over religious differences.

Mark Twain was similarly caught in a severe storm. He, too, made promises to God. However, the next day was bright and sunny and...well, he forgot his promises.

# 20 Lightning Protection

## Lightning Protection

We can learn to lower the risk of death, injury or property damage from lightning. There are many rules that one can follow for lightning safety, but the study of lightning is too recent for rules to have found their way into the form of proverbs.

## Zones of Protection

There is a rule of thumb that if you draw a circle on the ground with a radius equal to the height of the tall object, all strikes that would reach the level surface will go to the tall object. Around this tall object there is a *zone of protection.*

However, this rule only holds if the tall object is *well grounded.* When the object is not well grounded, there is *no* zone of protection. We use this rule when lightning rods are

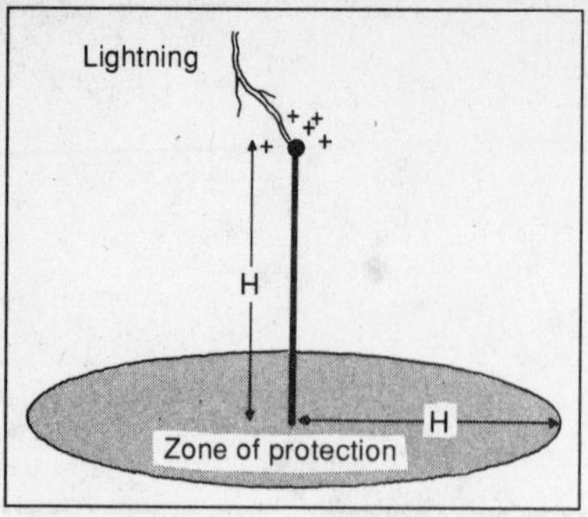


All lightning strikes that would reach a circle with a radius equal to the height of a grounded tall object will go to the tall object.

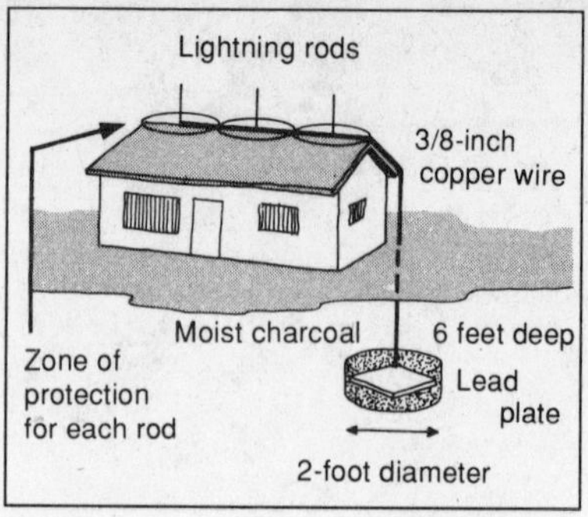


Overlapping circles of protection create zone of safety covering the building. Notice the good ground connecting all of the lightning rods.

placed on a building. Enough vertical rods are placed along a high ridge of the building to collectively create a zone of protection to cover the building.

The rods are then connected together and to a good ground by a 3/8-inch copper cable capable of carrying the large lightning currents. The problem with many installations is the nature of the good ground. It should be a heavy copper rod connected to a lead plate surrounded by wet charcoal about six feet under ground. The ideal installation will give a total resistance of less than one ohm. It can be very difficult to obtain a good ground in desert areas. In those cases, it may even be necessary to include a water-drip supply so the ground around the copper electrode is kept moist.

If the ground is not so good and has a resistance of 10 ohms, then a lightning strike to the system with a current of 100,000 amperes would raise the voltage of the rod system to one-million volts where breakdown would occur and allow the lightning current to enter other parts of the building.

## When Tall Objects Are Not Well Grounded

Many people killed by lightning are campers who have camped on the higher well-drained soil of a campground—

possibly near some tall trees. In this case, the lightning may strike a tall tree whose root system does not provide an adequate ground. The struck tree will then go to very high voltage and start breakdown streamers that run horizontally over the surface of the ground. These horizontal streamers can go up one leg and down the other of a standing camper or horizontally through the body if the person is lying down. In either case, the camper is electrocuted. Golfers also tend to run under a tree when the rain starts and can be killed or injured by these horizontal streamers.

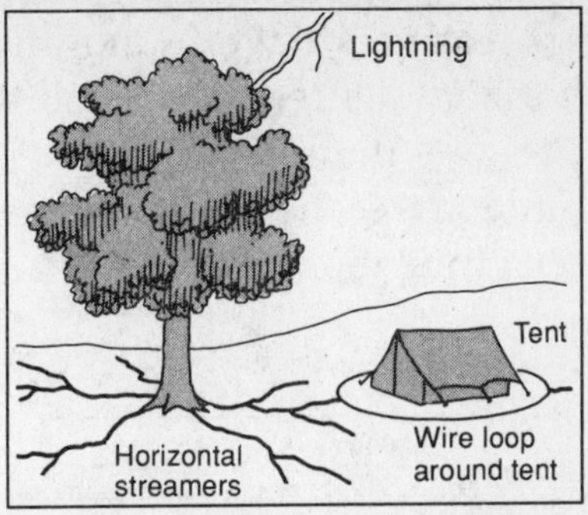


A closed loop of wire around the tent will direct horizontal lightning streamers away from the tent.

Oak trees have a deep tap root, thus giving them a good ground. Oak trees suffer more from direct lightning strokes, but there are fewer horizontal streamers from them during the strike. It is safer to be under an oak tree than under a birch tree which has a shallow root system.

Many people who are killed in this way could be revived if given proper attention after the strike. Any one horizontal streamer over the ground surface carries only a fraction of the total lightning current. If this fractional current finds its way through the body on a path which includes the heart, the heart will fibrillate (rapid quivering and spontaneous contractions of heart muscles). Death results if the victim is not resuscitated. If the current path is through the lower part of the brain, the autonomic breathing system fails and the person suffocates. A person struck by these horizontal streamers can often be revived when given proper heart massage and mouth-to-mouth resuscitation or CPR.

When camping, it is advisable to surround your tent with a closed loop of copper wire so that these horizontal streamers can follow the wire around the tent.

Farmers have found an entire herd of cattle killed by horizontal streamers when the cattle huddled under a tree during a thunderstorm. If a farmer finds his cattle huddling under a tree, he should connect the tree to a good ground the next time the weather is fair.

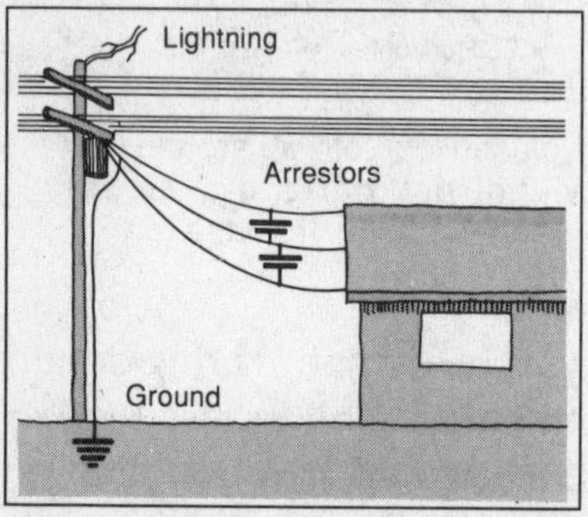


Typically three lines enter a dwelling or business. Two are at 115 volts and the third is at ground potential. The two hot lines provide 230 volts between them for stoves, dryers, etc. The arrestors are then placed between the hot lines and the power-line ground. No additional ground is required.

## How to Stay Safe Inside a Building

Many times lightning current finds its way into a building from a strike some distance away. Lightning currents tend to follow any conducting path to ground, such as a power line into a building, a telephone line, or a television cable. A good rule is to stay away from electrical appliances, television sets, plumbing fixtures, and telephones during a thunderstorm.

Another rule is to stay out of the bathtub because this places you in good contact with the water-pipe system, which is a favorite path for the current to ground once lightning current enters the building. When a conducting line enters a building, the entrance point should be fitted with a lightning arrestor. Engineers have devised these devices so that, when lightning currents cause an overvoltage on the line, the arrestor takes on a low resistance and shorts the excess currents to ground. Once the overvoltage vanishes, the lightning arrestor recovers to a large resistance.

## Lightning on Water

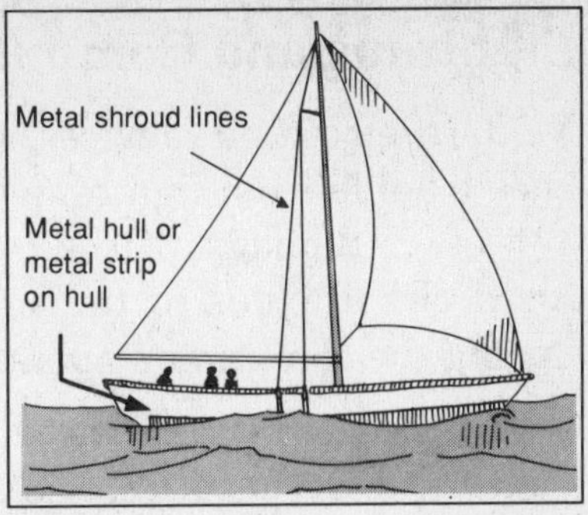

Metal shroud lines should be attached to the grounded hull on the outside of the boat to redirect lightning strikes.

Boats—especially sailboats—are always the highest point on a flat surface of water and enhance the probability for being struck by lightning. If there is a metal mast, it should be connected to a metal strip that makes good contact with the water along the hull. Any shroud line that supports the mast should be of metal and connected to the grounded hull over the side of the boat. The navies of the world powers in sailing-ship days lost more ships to lightning than from enemy engagements! Not only would the lightning shatter the mast, but the lightning currents would also explode the powder magazines. This problem was greatly reduced when a strip of copper was nailed along the wooden mast and extended over the side into the water.

It is also dangerous to be swimming, especially in fresh water, during a thunderstorm. Fresh water is a sufficiently poor conductor so that horizontal streamers develop and spread the currents over a large surface area. Salt water is a sufficiently good conductor to allow lightning currents to penetrate into the water.

## A Safe Place During a Thunderstorm

A safe place to be is inside a modern building with many steel girders used in its construction. These steel pieces serve as good conductors to a good ground. The building may be struck, but the currents are all conducted safely to ground. A reasonably safe place to be is inside a steel-bodied automobile. The lightning will follow the outside steel shell and arc across the tires to ground. The tires do not insulate the car from ground during a lightning strike.

## Lightning and Computers

Modern electronic devices, such as computers, can experience trouble from lightning even if there is no indication of any other damage. These troubles can be internally punctured circuit chips or transmission of an incorrect message during the computer's operation.

There are lines of magnetic force which always surround a current-carrying wire. The magnitude of the magnetic force is proportional to the current in the wire. For lightning currents, these lines of force can become quite intense—even at large distances from the lightning current. In the case of lightning currents, the intensity changes very fast with time. If these lines of magnetic force can thread through another circuit, they will induce extraneous voltages in that circuit. This voltage is proportional to the rate of change of the magnetic field and the area of the threaded circuit.

Computer design engineers try to keep the circuit areas small, yet there are always some areas that can be threaded by the magnetic lines of force of lightning current. Many computer circuits operate at only five volts. The extraneous induced voltages can easily exceed the five-volt rating of circuit components and this leads to trouble.

Even if the computer is in a building well protected from lightning, the computer can still get into trouble if it is located close to lightning currents on their way to ground. All along the way we have arrived at lightning protection by providing good paths to ground for the lightning current. If the computer is near some steel beam in a well-constructed building, that beam can carry large lightning currents to ground. And the associated lines of force can still get to the computer. In critical situations, the computer circuits can be protected by enclosing the computer in a good conducting copper case which diverts the magnetic lines of force around the case. But this is inconvenient and expensive.

Because the computers are operated on electric power, much of the trouble can enter the computer through the power lines when the building has no lightning arrestor. If computers are used extensively, they should all be well grounded with an *additional ground*. The power-line ground is not adequate. The ground can be improved by tying all computer chassis to a heavy copper wire, which, in turn, is connected to a point where the water main enters the building. This should preferably be in the basement if there is one. The building power input should also be fitted with one or more lightning arrestors.

## Lightning Proverb

In most of the cases, I have made protection from lightning seem like a life and death matter. In spite of all the safety rules previously given, none are found in proverb form. Because I could find few good proverbs to cover lightning, I took the opportunity to make a new one to remind us of things we can do for our safety.

***When thunderstorms come our direction,***
***One should think of good protection;***
***There are zones where this is found,***
***Be sure the lightning gets to ground.***

## Invasion of a Thunderstorm

You may think the proverbs in this book don't give sufficient information to prevent getting caught in a thunderstorm after you predicted fair weather. But this shouldn't happen very often if you learn some basic meteorology principles and stay alert.

Although your area may need precipitation to make agriculture successful and to wash pollution from the air, you may be disappointed if a large storm comes when you'd planned some fair-weather activity. So be prepared to make the best of any weather situation.

A severe thunderstorm upsets nearly everybody's routine,

even though one may be in a protected building. The chances are finite that, through some strange behavior of lightning, a family friend who does not take necessary precautions may not survive. The Greeks knew this well and believed that Zeus, the Chief of the Gods, was reckoning with man for his transgressions. It may be that your fears stem from such beliefs.

The Judaeo-Christian God exercised his will with severe storms as told early in the *Bible.* Exodus 9:23 states, "And Moses stretched forth his rod toward Heaven; the Lord sent forth thunder and hail and the fire ran along the ground." Perhaps you think reverently and silently offer a prayer to spare yourself from direct interaction with a thunderbolt during a storm. When lightning strikes and does harm, you think there are so many more places it could have struck without doing harm. After your first fears, you try to go about things as usual if you can.

If you like to observe lightening storms, there is a lot to be learned. Some pointers follow.

## Viewing a Lightning Display

Many of us are fascinated by lightning displays and simply like to watch the ribbons of fire dance through the sky as a thunderstorm approaches. Although the best view is from an open area where you can see in all directions, this is a terribly unsafe place to be because you can be the tallest object in the region around you.

You should *always* be surrounded by some form of a conducting metal structure during a thunderstorm. This is called a *Faraday Cage.* A safe place to be is at a window in a high-rise apartment building where the steel construction beams form such a protective cage. But this only lets you see from one side of the building—unless it is a corner window. You'll miss what happens on the other sides of the building. A home with a set of lightning rods can serve equally well.

You can safely view the storm in an open area from inside a steel-roofed automobile, van or truck. Just keep your hands off of the metal body parts. If lightning strikes the vehicle, you'll have a memorably frightening experience because of the bright flash and loud noise. The lightning currents are carried around the viewer by the metal car body and find their way to ground by arcing across the tires. Being outside the car is one of the *unsafest* places to be.

Many people like to view storms from a roofed porch which gives protection when the storm is 10 or more miles away. I know of no cases where people have been injured in this way. If you count the number of seconds between the flash and the thunder (as described below), this form of viewing can be quite safe. As soon as the time between the flash and the thunder becomes less than 10 seconds (2 miles away), immediately move inside the house where water-pipes and electrical power wires form a cage of conducting materials around you. When you are inside the house, don't touch any appliance connected to this shield system such as a stove, a telephone, any plumbing fixture or bath water.

## Determining Distance and Length of a Channel

Since we see the light of a lightening flash almost instantaneously and sound from the heated channel travels at about 1100 feet per second (about 1 mile per 5 seconds), we can determine the distance to the flash as well as estimate the length of the channel.

To determine how far away a flash was, start counting seconds—one thousand one, one thousand two, one thousand three, and so on—immediately upon seeing the flash until you hear thunder. Note the seconds. On hearing the first thunder, note the number of seconds of the first count, but immediately start a second count. Continue counting as long as you can hear the thunder. The first count will tell you how far away the closest part of the flash was. Allow five

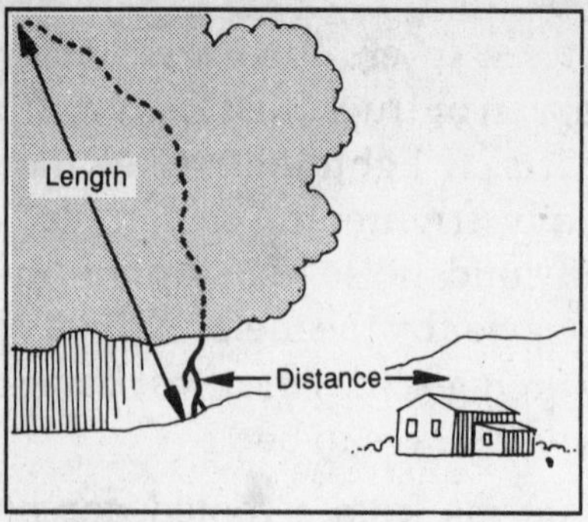


You are one mile away from lightning for every five seconds it takes from the time you sight it until you hear the sound of thunder. Length of the lightning channel is at least one mile long for every five seconds of thunder.

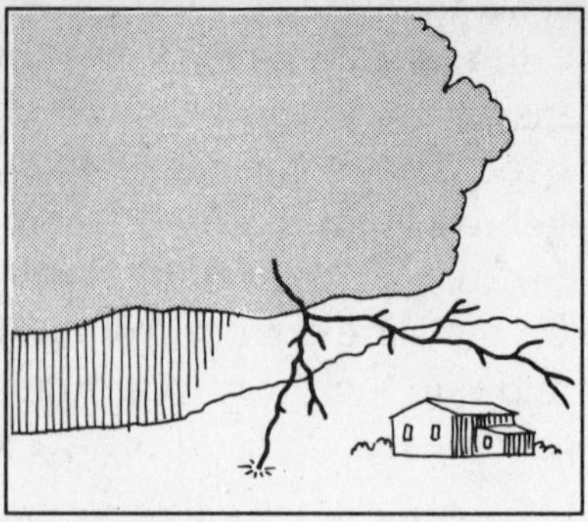

A tearing or ripping sound prior to hearing the first crack of lightning indicates a branch has gotten closer to you.

seconds for each mile. For example, if you got to one thousand thirteen on the first count, dividing 13 by five makes the flash about two and one-half miles away. The second count follows the arrival of thunder from distant regions along the channel. This may go to one thousand thirty, telling you the channel is at least six-miles long.

The phenomenon is much like forming a column of soldiers with rifles strung in a line for several miles. On a given visual command they all fire their rifles at the same time. The sound arrives from the closest soldier first, then continues to arrive from progressively distant soldiers. If the column had many kinks and bends, you would hear a rumble and the pitch would get lower and lower as high frequencies become absorbed by water vapor in the air.

This counting technique will tell you the distance of the closest soldier, but your estimation of the length of the column depends on its orientation with respect to you. The column will be at least as long as your second count, but may be much longer. In this second count you can get some estimate of the size of a storm cloud. If the second count is very long, you are dealing with a big storm.

Many lightning flashes are between opposite charges

within clouds and do not reach the ground. Counting techniques can still be used, so you get some idea of the lowest point in the cloud and extent of the cloud. Depending on how sensitive your ear is to changing pitch, the rumble of thunder can also tell you how far away things are happening.

Prior to the first crack of a close flash, you sometimes hear a tearing or ripping sound. This is from one of the branches that has gotten closer to you than the main stroke, which reached the earth somewhere else. When you hear this noise, feel fortunate that it wasn't that particular branch of the stroke that found its way to ground.

## Photographing Lightning

You may want to get some pictures of lightning. To do this successfully the flashes should be close or you should use a telephoto lens so that most of the picture is filled with lightning phenomena. Pictures are taken by using the *time exposure* or *bulb* setting on your camera. Keep the camera steady by mounting it on a tripod.

This means that pictures must be taken only at night and away from city lights so the background will not be overexposed. You must guess where the next flash will occur and point the camera in that direction. Use maximum lens opening to get more detail.

Open the shutter, wait for the flash, then close the shutter immediately after the flash. Background lighting is usually such that the exposure should be limited to a one- or two-minute duration. If you don't get a flash in this time, start a new exposure. It is disappointing to get a beautiful lightning-flash picture that's been spoiled by excess background exposure.

A storm usually lasts for about 20 minutes, so a 20-exposure roll of film should cover one storm period.

For a more sophisticated camera technique, *pan* the flash with the camera—slowly move the camera across the sky with the shutter open—so multiple strokes in the flash can be resolved. Each stroke will show up as a separate image. Only the first image will have branches. You can compare the number of images with the observed flicker in the flash.

## Questions About Lightning?

If you have any questions about lightning, please send them with a stamped, self-addressed envelope to Dr. George Freier, University of Minnesota, School of Physics and Astronomy, 116 Church Street, S.E., Minneapolis, MN 55455.

## Had An Experience With Ball Lightning?

If you have had a personal experience with ball lightning (described on page 191), I would very much like to hear about it. Please write to me with a description of the ball size, color, duration, sound, smell, damage, or any other feature of the event. You can send it to the above address.

# Index

## D

## I–K



## K



## L



## M

## N



## O



## P

## R



## S

## T



## U–V



## W

## Z